Learning Materials in Biosciences

Learning Materials in Biosciences textbooks compactly and concisely discuss a specific biological, biomedical, biochemical, bioengineering or cell biologic topic. The textbooks in this series are based on lectures for upper-level undergraduates, master's and graduate students, presented and written by authoritative figures in the field at leading universities around the globe.

The titles are organized to guide the reader to a deeper understanding of the concepts covered.

Each textbook provides readers with fundamental insights into the subject and prepares them to independently pursue further thinking and research on the topic. Colored figures, step-by-step protocols and take-home messages offer an accessible approach to learning and understanding.

In addition to being designed to benefit students, Learning Materials textbooks represent a valuable tool for lecturers and teachers, helping them to prepare their own respective coursework.

Kota Miura · Nataša Sladoje
Editors

Bioimage Data Analysis Workflows — Advanced Components and Methods

 Springer

Editors
Kota Miura
Nikon Imaging Center
University of Heidelberg
Heidelberg, Germany

Nataša Sladoje
Department of Information Technology
Uppsala University
Uppsala, Sweden

ISSN 2509-6125   ISSN 2509-6133   (electronic)
Learning Materials in Biosciences
ISBN 978-3-030-76393-0   ISBN 978-3-030-76394-7   (eBook)
https://doi.org/10.1007/978-3-030-76394-7

This Springer imprint is published by the registered company Springer Nature Switzerland AG.
The registered company address is: Gewerbestrasse 11, 6330 Cham, Switzerland

Acknowledgement

This textbook is based upon the work from COST Action CA15124 NEUBIAS, supported by COST (European Cooperation in Science and Technology).

COST (European Cooperation in Science and Technology) is a funding agency for research and innovation networks. Our Actions help connect research initiatives across Europe and enable scientists to grow their ideas by sharing them with their peers. This boosts their research, career, and innovation.

► www.cost.eu

Contents

Editors and Contributors

About the Editors

Kota Miura Nikon Imaging Center, University of Heidelberg, Heidelberg, Germany
e-mail: kota.miura@gmail.com

Nataša Sladoje Centre for Image Analysis, Department of Information Technology, Uppsala University, Uppsala, Sweden
e-mail: natasa.sladoje@it.uu.se

Contributors

Ignacio Arganda-Carreras Departamento de Ciencias de la Computación e Inteligencia Artificial, Facultad de Informática, Universidad del Pais Vasco, Guipúzcoa, Spain
e-mail: ignacio.arganda@ehu.eus

Arianne Bercowsky Rama EPFL SV IBI-SV UPOATES AI 3133 (Bâtiment AI) Station 19, Lausanne, Switzerland
e-mail: arianne.bercowskyrama@epfl.ch

Bertrand Cinquin UMS3750, IPGG – Institut Pierre Gilles de Gennes, Paris, France
e-mail: bertrand.cinquin@espci.fr

Daniel Franco-Barranco Donostia International Physics Center (DIPC), Guipúzcoa, Spain
e-mail: daniel.franco@dipc.org

Estibaliz Gómez-de-Mariscal Department of Bioengineering and Aerospace Engineering, Biomedical Imaging and Instrumentation Group, Universidad Carlos III de Madrid, Leganés, Spain
e-mail: esgomezm@pa.uc3m.es

Robert Haase DFG Cluster of Excellence "Physics of Life", Technische Universität Dresden, Dresden, Germany
e-mail: robert.haase@tu-dresden.de

Joyce Y. Kao Department of Computer Science, Swiss Federal Institute of Technology (ETH) Zürich, Zürich, Switzerland
e-mail: joyceykao@gmail.com

Anna Klemm SciLifeLab BioImage Informatics Facility and Department of Information Technology, Uppsala University, Uppsala, Sweden
e-mail: anna.klemm@it.uu.se

Marion Louveaux Institut Pasteur, Paris, France
e-mail: marion.louveaux@pasteur.fr

Arrate Muñoz-Barrutia Department of Bioengineering and Aerospace Engineering, Biomedical Imaging and Instrumentation Group, Universidad Carlos III de Madrid, Leganés, Spain
e-mail: mamunozb@ing.uc3m.es

Mark L. Siegal Center for Genomics and Systems Biology, Department of Biology, New York University, New York, NY, USA
e-mail: mark.siegal@nyu.edu

Stephane Verger Umeå Plant Science Center, Department of Forest Genetics and Plant Physiology, Swedish University of Agricultural Sciences, Umeå, Sweden
e-mail: stephane.verger@slu.se

Daniela Vorkel Center for Systems Biology Dresden, Max Planck Institute for Molecular Cell Biology and Genetics, Dresden, Germany
e-mail: vorkel@mpi-cbg.de

Yishaia Zabary Department of Software and Information Systems Engineering, Ben-Gurion University of the Negev, Beersheba, Israel
e-mail: yshaayaz@gmail.com

Assaf Zaritsky Department of Software and Information Systems Engineering, Ben-Gurion University of the Negev, Beersheba, Israel
e-mail: assafzar@gmail.com

Reviewers

- Chapter 2: Jan Eglinger, FMI, Basel, Switzerland
- Chapter 3: Uwe Schmidt, Center for Systems Biology MPI-CBG, Dresden, Germany
- Chapter 3: Martin Weigert, EPFL, Lausanne, Switzerland
- Chapter 4: Sébastien Tosi, IRB Barcelona, Barcelona, Spain
- Chapter 5: Dominic Waithe, University of Oxford, Oxford, UK
- Chapter 6: Mafalda Sousa, I3S – Advanced Light Microscopy, University of Porto, Porto, Portugal
- Chapter 7: Jonas Øgaard, Research Institute of Internal Medicine, Oslo University Hospital, Oslo, Norway
- Chapter 8: Simon F. Nørrelykke, ETH Zurich, Zürich, Switzerland

Introduction

Nataša Sladoje and Kota Miura

Contents

© The Author(s) 2022
K. Miura, N. Sladoje (eds.), *Bioimage Data Analysis Workflows–Advanced Components and Methods*,
Learning Materials in Biosciences, https://doi.org/10.1007/978-3-030-76394-7_1

1.1 Introduction

Bioimage analysis is often regarded as a technical task that can be solved simply by the development of new and more sophisticated image processing algorithms. This may be true to a large extent, but the complexity encountered in the actual usage of those algorithms during the analysis leads to a number of challenges that leave researchers with a thought that *"Bioimage analysis is difficult"*.

To provide structure and organization in this complexity, and by that to enable and simplify users' navigation through it, the Network of European Bioimage Analysts (NEUBIAS) has been systematically looking at the computational tools and algorithmic resources of bioimage analysis with a slightly higher resolution, identifying components, collections, and workflows (Miura et al., 2020). Each of these terms define different types of computational tools. **Component** is an implementation of a certain image processing or analysis algorithm; **Collection** is a software package, or a library, that includes many valuable (possibly independent) components, and is offered as a collection of downloadable files, ready to be used; **Workflow** is created by assembling the components (e.g., from one or more collections) into a sequence of image processing and analysis steps, to solve a certain biological question. Workflows typically take raw image data as input and aim at delivering parameters of biological systems and/or visualization of the system analysis results as an output.

For creating a workflow, knowledge of the characteristics of various components and their behavior against image data is required. At the same time, one needs to know some standard methods for assembling components into workflows. Furthermore, the ability of a user to use programming language becomes mandatory, as it dramatically enhances the range of components one can select from, and increases the efficiency of automated analysis. Moreover, presenting a workflow in the form of a computer program can be regarded as a highly recommended scientific practice for method reproducibility. Therefore, the training in bioimage analysis should ideally include the three main elements: component-related literacy, programming language fluency, and workflow design.

In the previous textbook (Miura and Sladoje, 2020) prepared by, and for, the NEUBIAS community, as well as in the earlier BIAS textbook (Miura, 2016b), we focused primarily on introducing the main principles of workflow design, and how to implement workflows using scripting languages such as ImageJ Macro, MAT-LAB, and R. We selected this particular approach with an aim to reduce the imbalance between the vast amount of already existing literature and textbooks focused on image processing and analysis algorithms (i.e. components), in comparison with scarce resources for learning how to design and implement bioimage analysis workflows. Contributing authors were asked to provide a holistic view of a bioimage analysis task, starting with introducing the biological background relevant for their chapter, and describing the biological research question that they want to address by a fully-coded and reproducible image analysis workflow. In addition, they were expected to provide a detailed explanation of the code, and – finally – interpretation of the results of the performed analysis, in terms of the biological question in focus. These contributions narrowed the gap, at least to some extent as we believe, between the ever growing number, excellence, and complexity of image analysis components on one side, and the biological questions to be addressed by them, on the other side.

The textbook was warmly endorsed by the community of life scientists, as well as bioimage analysts, who have been using it as a valuable resource.

The Network of European Bioimage Analysts has been continuously growing, in terms of size and competence. Via education and communication, supported by numerous training sessions, dedicated conferences, research collaborations, discussion forums and several other activities, our members have learned many of the techniques used in the highly multidisciplinary field of bioimage analysis. To best respond to their needs, we have decided to widen the scope of this new Bioimage Analysis textbook in two ways.

Firstly, we have included several chapters devoted to **components**, in response to the increasing demand to use the cutting-edge algorithms and follow the most recent trends in the field of bioimage analysis. In our opinion, this demand is a direct consequence of the narrowed gap in interdisciplinary competences and communication between life scientists (biologists) and computer scientists, and increased competences of bioimage analysts who have become more skilled in bridging those two fields. We appreciate this as a valuable outcome of various efforts made by the NEUBIAS community: Increased interest and utilization of high-end components in life sciences result from the presence of the new experts. Note that, however, the authors of these "component" chapters preserved the main flavour of our textbooks – the biological context, exemplified by use-cases of the presented components, possibly within workflows.

Secondly, the increase in the number of bioimage analysts, but also in the level of their skills and competences, has motivated us to include another novel form of contributions: "**workflow deconstruction**" chapters. This pedagogical approach in bioimage analysis training has been proposed by Jean-Yves Tinevez and Kota Miura, inspired by the "Deconstruction" concept introduced by postmodern philosopher Jaques Derrida, and developed during the NEUBIAS Training Schools for Bioimage Analysts in several editions between years 2016 and 2020. The aim has been to maximize the learning experience in workflow design by learning to generalize the knowledge and techniques gained from a small sample of well selected examples of workflows, deconstructed and discussed in detail with respect to how components are assembled and critically evaluated within the design. This approach was suitable for the trainees proficient in computer programming and experienced in usage of a variety of components; these were primarily professional bioimage analysts. We hope that "workflow deconstruction" chapters included in this book provide insight in the essence of workflow design, and also ignite readers' creativity in suggesting their own novel bioimage analysis workflows.

As a result, this Volume 2 collection includes seven chapters. The book starts with discussion on "*Batch Processing Methods in ImageJ*" (▶ Chap. 1), and presentation of tools available in "*Python: Data handling, analysis and plotting*" (▶ Chap. 2), both aiming to increase the fluency in programming languages, "tidy" data handling, and environments widely used in bioimage analysis. The subsequent chapters are focused on components: "*Building a Bioimage Analysis Workflow using Deep Learning*" (▶ Chap. 3) and "*GPU-accelerating ImageJ Macro image processing workflows using CLIJ*" (▶ Chap. 4); both describe ways to include cutting-edge components into a variety of workflows, responding to clear demands from the bioimage analysis community. We continue, and conclude, with three chapters devoted to workflow deconstruction, putting in focus three different biological problems, and suggesting and analysing their original suggested solutions: *"SurfCut macro deconstruction"* (► Chap. 5), *"i.2.i. with the (fruit) fly: Quantifying position effect variegation in Drosophila melanogaster"* (► Chap. 6), and *"A MATLAB pipeline for spatiotemporal quantification of monolayer cell migration"* (► Chap. 7). These chapters require certain literacy in programming, but offer numerous valuable tips out of which many are generally applicable. In particular, ► Chap. 5 aims to provide an introduction to the concept and practice of "workflow deconstruction", demonstrating the process in detail. ► Chapters 6 and 7 follow with examples of very successful original designs and utilization of image analysis workflows to perform detailed and unique analysis of extracted biological parameters.

► Chapters 1, 3, 4, 5, and 6 require some basic knowledge of ImageJ macro language. If lacking it, the readers are referred to "ImageJ Macro Language" (Miura, 2016a). ► Chapters 2 and 3 assume basic knowledge of Python programming. ► Chapter 7 requires basic knowledge of MATLAB programming. There are numerous available resources to support readers to meet these requirements; in particular, we mention "Introduction to MATLAB" (Monzel and Möhl, 2016) and "Introduction to MATLAB" (Nørrelykke, 2020), the former being general and basic, and the latter slightly more advanced.

This textbook is the 2nd bioimage analysis textbook published as an output of the common efforts of NEUBIAS, funded under COST Action CA15124. We would like to thank the project workgroup (WG) leaders: Sebastian Munck, Arne Seitz, and Florian Levet (WG1 "Strategy"); Paula Sampaio and Irene Fondón (WG2 "Outreach"); Gaby Martins and Fabrice Cordeliéres (WG3 "Training); Perrine Paul-Gilloteaux and Chong Zhang (WG4 "Webtool biii.eu"); Sébastien Tosi, Graeme Ball and Raphaël Marée (WG5 "Benchmarking and Sample Datasets"); Julia Fernandez-Rodriguez and Clara Prats Gavalda (WG7 "Short-Term Scientific Missions and Career Path"); and Julien Colombelli (the Action Chair). Their efforts to create a synergistic effect of the diverse workgroup activities towards the establishment of "Bioimage Analysts" is the strong backbone that has led to the successful realization of this book as a result of WG6 "Open Publication" (led by Editors). We are very much grateful to the reviewers of each chapter: Jan Eglinger, Uwe Schmidt, Martin Weigert, Sébastien Tosi, Dominic Waithe, Jonas Øgaard, Mafalda Sousa, and Simon F. Nørrelykke. Their critical comments largely improved the presented content. We are particularly grateful to the authors of each chapter: Anna Klemm, Kota Miura, Arianne Bercowsky Rama, Estibaliz Gomez-de-Mariscal, Daniel Franco-Barranco, Arrate Muñoz-Barrutia, Ignacio Arganda-Carreras, Daniela Vorkel, Robert Haase, Bertrand Cinquin, Joyce Y. Kao, Mark L. Siegal, Marion Louveaux, Stephane Verger, Yishaia Zabary, and Assaf Zaritsky; for their selfless commitment to meet the demanding requirements of the publication format that we have chosen. The publication of this book was enabled by the financial support from the COST Association (funded through EU framework Horizon2020), through the granted project "A New Network of European Bioimage Analysts (NEUBIAS, COST Action CA15124)". Finally, we wish to thank all members of NEUBIAS who, with their enthusiasm and commitment to the network's diverse activities, have contributed to keep the momentum of the initiative constantly high, a vital element to enable it to reach its objectives, including the publication of this book.

References

Monzel C and Möhl C (2016) Introduction to MATLAB. In: 'Bioimage Data Analysis. Wiley-VCH, Weinheim, pp 63–97. https://analyticalscience.wiley.com/do/10.1002/was.00050003

Miura K (2016a) ImageJ Macro Language In: Miura K (ed) Biomage data analysis. Wiley-VCH, pp 19–62. https://analyticalscience.wiley.com/do/10.1002/was.00050003

Miura K (ed) (2016b) Bioimage data analysis. Wiley-VCH, Weinheim. https://analyticalscience.wiley.com/do/10.1002/was.00050003

Miura K, Paul-Gilloteaux P, Tosi S, Colombelli J (2020) Workflows and Components of bioimage analysis. In: Miura K, Sladoje N (eds) Bioimage data analysis workflows. Learning materials in biosciences. Springer International Publishing, Cham, pp 1–7. https://doi.org/10.1007/978-3-030-22386-1_1

Miura K, Sladoje N (eds) (2020) Bioimage data analysis workflows. Learning materials in biosciences, Springer. OCLC: 1127266601. https://doi.org/10.1007/978-3-030-22386-1

Nørrelykke SF (2020) Introduction to MATLAB. In: Miura K, Sladoje N (eds) Bioimage data analysis workflows. Learning materials in biosciences. Springer International Publishing, Cham, pp 97–141. https://doi.org/10.1007/978-3-030-22386-1_5

Batch Processing Methods in ImageJ

Anna Klemm and Kota Miura

Contents

This Chapter has been reviewed by Jan Eglinger, FMI Basel, Switzerland.

© The Author(s) 2022
K. Miura, N. Sladoje (eds.), *Bioimage Data Analysis Workflows–Advanced Components and Methods*,
Learning Materials in Biosciences, https://doi.org/10.1007/978-3-030-76394-7_2

References – 27

What You Will Learn in This Chapter

In this chapter you will learn how to execute a workflow on not only one image but on several images in ImageJ – a technique that is called "Batch Processing". Various ways of doing this are possible in the Fiji distribution of ImageJ, and the characteristics of each and how-to are explained.[1]

2.1 Introduction

As many people may think, the most prominent power of a computer is the automation. This is also the case with image analysis in life sciences: the automation of bioimage analysis. While many tasks can be done fully manually, those workloads of analysis become much less by automating some of those tasks. Moreover, you will be able to scale up the number of analysis results by automation, which leads to more reliable results. Finally, the automation of bioimage analysis avoids human errors. The probability of the occurrence of human errors increases as manual working time increases, but with automated processing, this does not happen.

Batch Processing is a way of automation. With this technique, many images are processed one-by-one by repeated iteration of image loading, analyzing, and saving of the results. In addition to the advantage of automating the analysis, it allows to economize the usage of computer memory as only a single image is analysed per loop.

2.2 Types of Batch Processing Methods in ImageJ

With Fiji, there are multiple ways to do batch processing and analysis (Schindelin et al., 2012; Schneider et al., 2012). They are redundant with their goal, but are different in the way they are designed and used. Each method is optimized for certain usage, and it is good to know all of them so that you can select a suitable method depending on the situation.

The GUI-based method is convenient if you are unsure about your capability in writing macros. You just need to acquire macro commands using the Command Recorder, and copy & paste those commands in the GUI (see ▶ Sect. 2.6). This method is also good when you need to quickly document and provide information how to process many images by batch processing. One weak point of the GUI method is that it does not allow you to customize the saving of images and analysis results. We will see this later.

Though the GUI-based method is easy to use, the scripting based method of batch processing is more flexible to customize. If you know how to run *for*-loops with ImageJ macro, to include batch processing might be quite an easy job. In that case you just need to know how to handle file path operations and file naming (see ▶ Sect. 2.7.2). Moreover, by using an ImageJ2 functionality called "script parameters" (Rueden et al., 2017), it is possible to further enhance the generality of the scripts (see ▶ Sect. 2.7.3). This allows you to use the "batch" button in the Script

---

[1] This chapter was communicated by Jan Eglinger, FMI Basel, Switzerland

Editor (see ► Sect. 2.7.4), or to run the macro from command line in "headless" mode (see ► Sect. 2.7.5).

Finally, we will explore how to handle collecting data during batch processing (see ► Sect. 2.8), and demonstrate a simple application example of batch processing for data inspection (see ► Sect. 2.9).

2.3 Tools

— Fiji
  – Download URL: ► https://imagej.net/Fiji/Downloads
— Command Line Interface
  – Windows: Install Git for Windows. It comes with a BASH terminal (Git BASH).
    – Download: https://gitforwindows.org/
  – Mac: Terminal.app comes as default with the OSX system and can be used as it is.

2.4 Dataset

The dataset that we will work on to demonstrate batch processing consists of 3-channel images (16-bit) of the HeLa cells. Channel 1 represents the microtubules, Channel 2 shows a GFP-labelled nuclear protein, and Channel 3 contains the nuclei labelled with the marker DAPI.

For downloading codes and sample image data used, please access the following repository:

  ► https://github.com/NEUBIAS/neubias-springer-book-2021

All files that appear in this chapter are freely downloadable from there.

2.5 Core Workflow for Processing a Single Image

Our task here is to analyze a large number of images (in the provided dataset) and get results for each of them − i.e., to perform batch processing and analysis. But before making the batch processing workflow, we need to define the main part of the processing, the steps which are to be done for each single image.

The ImageJ macro file, which you can find in the code repository,[2] is a simple workflow that runs on an open, active single image. It segments the nuclei (Channel 3, C3) by setting a global automatic threshold, gets the outlines of the single nuclei, and measures the area, in pixels, of each nucleus. Let us see the actual code. There are roughly three steps.

— line 11–12: Duplication of C3 (Nuclei) to isolate the image for further analysis

---

2 01_Basic_Workflow_raw.ijm

```
11   Stack.setChannel(3);
12   run("Duplicate...", "title=C3_" + title); //Duplicate
     ↪   only C3 for further processing
```

01_Basic_Workflow_raw.ijm

— line 14–20: Filtering C3 and converting it to a binary image for segmentation

```
14   //median filtering to smoothen the image
15   run("Median...", "radius=10");
16   //set an auto threshold and binarize
17   setAutoThreshold("Li dark");
18   setOption("BlackBackground", true);
19   run("Convert to Mask");
20   run("Fill Holes");
```

01_Basic_Workflow_raw.ijm

— Line 23: Measuring the area of each nucleus and generating a Results Table, and the ROI Manager listing the outlines of each nucleus. Nuclei that are touching the border of the image are excluded from the measurements.

```
23   run("Analyze Particles...", "size=1000-Infinity
     ↪   display exclude add");
```

This macro[3] does not save any output (binary image, a window with the results, ROI Manager; see ◻ Fig. 2.1). They are only visible on the desktop after running the macro. However, it is our aim to save the binary image and the ROI Manager (both as quality control output), as well as the measurement results of each image. We will introduce the saving techniques as we explain each batch processing method below.

? Exercise 1

Go through the simple workflow macro 01_Basic_Workflow_raw.ijm line-by-line, analyse each command and make sure to understand what each of them is doing.

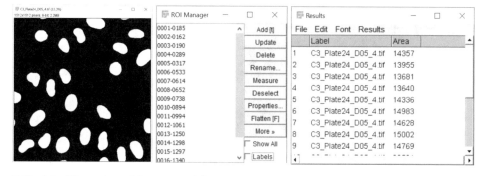

◻ **Fig. 2.1**   The output of the core workflow

---

3   01_Basic_Workflow_raw.ijm

2.6 GUI-Based Methods

A very easy method to execute code on an entire set of images is to use the so-called Batch Processor. The Batch Processor is a GUI found under [Process > Batch > Macro...].

Let us have a look on the settings of the Batch Processor (◘ Fig. 2.2).

1. Set input and output folder. Input is the folder where you store the images to be processed. Output is where the Processor will automatically store output images. Enter a path for both input and output directories or choose the output via dialog windows by clicking on ''Input...''/''Output...''.
2. Specify the format for saving the images by the drop-down menu of ''Output format''.
3. Within the large text field, enter the code to be executed on the images by either copy&pasting to the window, or by opening the simple workflow macro.[4]
4. Click the button ''Process'' to run the code on all images.

While running the batch-processor, we can observe that no image is opened in display; images are processed in the background. Regarding the output, we can find our binary control-images in the output folder, together with one big Results Table containing the measurements of all nuclei of all the (4) images we have worked with. However, ROIs that were in the ROI Manager were not saved.

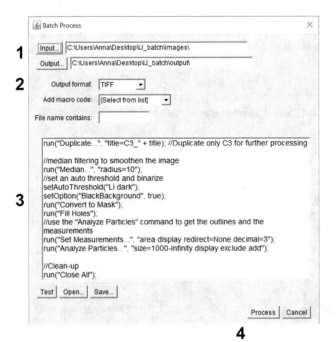

◘ **Fig. 2.2**   Batch Processor Interface

---

4   `01_Basic_Workflow_raw.ijm`

This nicely shows the power but also the disadvantages of the Batch Processor: it is very fast to run code on an entire set of images, however we do not have full control over what is saved (in this case, the ROIs were not saved). Also, if we would decide to save the, e.g., median filtered image as quality control, this would not be possible. Another disadvantage of the Batch Processor is that it does not handle subfolders, but only processes images on the level of the selected input folder.

2.7 Scripting-Based Methods

In order to have full control over the batch-processing, we need to write all the steps into our full IJ macro script.

2.7.1 Preparing the Code for Batch Processing

Before we execute our code on an entire set of images we want to make sure that the code includes commands to "clean-up" the traces of processing and analysis after each processed image.[5]

- We want to make sure that the ROI Manager does not contain ROIs of another image.
- Depending on the situation we can collect all results in one table or create and save one Results Table for each image. In the latter case we need to clean the Results Table between images.
- We want to close all open images after we analyzed one of the input-images, in order to have a fresh start for the next input image.

In the IJ macro language we can ensure doing so using the following commands for clearing the ROI Manager and Results Table:[6]

```
7    roiManager("reset");
8    run("Clear Results");
```

To close all images in the end, we can run:

```
30   run("Close All");
```

With these "clean-ups", we are now ready.

2.7.2 ImageJ Macro, IJ1

Now that we have prepared the code for the core workflow, we can work on the script to enable batch-processing. As a starting point for writing, a template of a batch-processing macro is available within the Script Editor under [Templates > ImageJ 1.x > Batch > Process Folder (IJ Macro)].

---

5   Ensuring a clean start is general good practice, not only when aiming for batch execution.
6   See 01b_Basic_Workflow_prepared.ijm

```
1   /*
2    * Macro template to process multiple images in a folder
3    */
4
5   #@ File (label = "Input directory", style = "directory") input
6   #@ File (label = "Output directory", style = "directory") output
7   #@ String (label = "File suffix", value = ".tif") suffix
8
9   // See also Process_Folder.py for a version of this code
10  // in the Python scripting language.
11
12  processFolder(input);
13
14  // function to scan folders/subfolders/files to find files with
    ↪   correct suffix
15  function processFolder(input) {
16          list = getFileList(input);
17          list = Array.sort(list);
18          for (i = 0; i < list.length; i++) {
19                  if(File.isDirectory(input + File.separator +
                      ↪  list[i]))
20                          processFolder(input + File.separator +
                            ↪  list[i]);
21                  if(endsWith(list[i], suffix))
22                          processFile(input, output, list[i]);
23          }
24  }
25
26  function processFile(input, output, file) {
27          // Do the processing here by adding your own code.
28          // Leave the print statements until things work, then remove
              ↪   them.
29          print("Processing: " + input + File.separator + file);
30          print("Saving to: " + output);
31  }
```

template_Process_Folder.ijm

The template contains three main sections:
- Line 5–7: Getting input folder, output folder and the type of the image to be analyzed;
- Line 15-24: processFolder function registers files to be analyzed and searches also in subfolders;
- Line 26–31: processFile function contains the core workflow processing the individual files.

Getting Input Folder, Output Folder

The first several lines of the template (template_Process_Folder.ijm) are utilizing so-called script parameters. For now, we will not use the script parameters and will instead hard-code input directory, output directory and the file suffix. The modified code is in another file, which we will call the macro with hard-coded path.[7] The script parameters have been commented out and instead input, output, and suffix were specified as follows:

```
10   input  = "C:/Users/Anna/Desktop/IJ_batch/images";
11   output = "C:/Users/Anna/Desktop/IJ_batch/output";
12   suffix = ".tiff";
```

02a_Process_Folder_Path.ijm

Let us make some remarks related to the path: (1) Make sure to adapt the paths to your local computer; (2) When copied from your system, the path contains either \ or /, depending on your operating system. If the path contains a backslash \ (older Windows OS) it is best to simply replace it by a slash /. If you want to use a \ you need to insert a second backslash: \\. This is called an escape character; (3) Instead of \ or / you can also use File.separator, which "Returns the file name separator character (/ or \)"[8] that is used by your system.

ProcessFolder Function

We have defined the input and output folders, and the suffix (file type). Let us now examine the processFolder function line-by-line. processFolder takes a path as argument. The path points to the folder with the input images, as defined by the user. First, in Line 16, getFileList acquires all the filenames in the directory as an array: each element of this array is a filename, and the length of the array is equal to the number of files within the directory. Array.sort(list) sorts this array in alphanumeric order.

Once we have the sorted list, we loop over it from Line 18, starting at $i = 0$ − since the index i of the first element of an array is 0 − until i is smaller than list.length, which gives the number of elements (= number of filenames) in list. For each filename, we create the full path of the file in Line 19 (input + File.separator + list[i]). Note again the usage of File.separator. Once the full path is created, we check whether the path is a directory using the command File.isDirectory. If yes, we call the processFolder function recursively in Line 20. In other words we now take the path of the detected directory as input and repeat the process of getting the FileList and checking each file of this new directory. Only if an item list[i] of our list of filenames is not a directory and endsWith the desired suffix, the function processFile is called. processFile in Line 22 will contain our core workflow and we can finally process and analyze the image. If a file is neither a directory, nor ends with the right file suffix, nothing happens.

❓ Exercise 2

Change the structure of your input folder. Create subfolders and subsubfolders and copy some of the images into them. Rename the images with e.g. "_level1". Run the Process_Folder.ijm template as it is and follow in which order the files are processed.

---

7   02a_Process_Folder_Path.ijm
8   ▶ https://imagej.net/ij/developer/macro/functions.html#File.separator

ProcessFile Function

The function `processFile` is executed on each file of the desired file type. Within the function, we want to first open an image, then run the core workflow to process and analyze the image, and finally save the output.

For opening a file, we use:

```
open(input + File.separator + file);
```

`file` is the filename stored in `list[i]` and passed on to the `processFile` function as the third argument: `processFile(input, output, file)`.

Until this point, we have modified `processFile` such that it opens an image. For processing the opened image, we can then copy & paste our simple workflow prepared for batch [9] within the function `processFile`.

Here you can see the first lines of our core workflow in the function:

```
28   function processFile(input, output, file) {
29       // Do the processing here by adding your own code.
30       // Leave the print statements until things work, then remove
         ↪  them.
31       print("Processing: " + input + File.separator + file);
32       print("Saving to: " + output);
33
34       //opening the image
35       open(input + File.separator + file);
36       filename_pure = File.nameWithoutExtension;
37       saving_prefix = output + File.separator + filename_pure;
38
39       //preparations
40       roiManager("reset");
41       run("Clear Results");
42       run("Set Scale...", "distance=0 known=0 pixel=1 unit=pixel");
         ↪   //we remove the scaling
```

Finally, we also want to save the output. Using scripting, we have full control over what to save. In the macro with hard-coded path, you can find the following strategy for saving:[10]

1. Line 36: Get the pure filename (the file name without the file extension) directly after opening the image. The function on the right-hand side gets "The name of the last file opened with the extension removed"[11].
2. Line 37: Create a saving prefix string that contains the output folder, the `File.separator` and the pure filename.
3. Lines 66–69:

```
66   saveAs("results", saving_prefix + "_results.csv"); //use saveAs
     ↪   command to save results
67   //save the isolated C3 (binary image)
68   selectWindow("C3_" + title);
69   saveAs("tiff", saving_prefix + "_C3.tif"); //use saveAs command to
     ↪   save an image
```

---

9 `01b_Basic_Workflow_prepared.ijm`

10 `02a_Process_Folder_Path.ijm`

11 ▶ https://imagej.net/developer/macro/functions.html#File.nameWithoutExtension

Save the Results window and the binary image, by using `saveAs(format, path)`. Let us read the documentation of the `saveAs` command:

> » Saves the active image, lookup table, selection, measurement results, selection XY coordinates or text window to the specified file path. The format argument must be "tiff", "jpeg", "gif", "zip", "raw", "avi", "bmp", "fits", "png", "pgm", "text image", "lut", "selection", "results", "xy Coordinates" or "text".

In consequence, for saving the Results window, we use "results" as format, and for saving an image as tif-file, we use "tiff" as format. Note that the active image is the one which is saved, so we need to make sure to activate the binary image by `selectWindow("C3_" + title);`.

4. Save ROIs in the ROI Manager. We use one of the `roiManager` functions:

```
71  roiManager("save", saving_prefix + "_rois.zip"); //drag&drop
    ↪  zip-file on Fiji to reopen ROIs
```

It creates a zip file. We can re-open this zip-file afterwards by drag&drop into Fiji: all ROIs will reappear within the ROI Manager.

We use the variable `saving_prefix` that we had built in the step before. It helps us to easily create the paths for the final output-file: we just need to add a suffix and file-ending, e.g. `_rois.zip`.

2.7.3 Two Different Methods to Get User Input

In the macro with hard-coded path,[12] paths and (file name) suffix are fixed.

```
10  input = "C:/Users/Anna/Desktop/IJ_batch/images";
11  output = "C:/Users/Anna/Desktop/IJ_batch/output";
12  suffix = ".tiff";
```

```
02a_Process_Folder_Path.ijm
```

This is useful e.g. when we develop a workflow and do not want to interactively select a file every time we run the code.

If we do want to allow for user input, we can devise graphical user interfaces, GUIs, as exemplified in the macro with user interface.[13]

```
9   input = getDirectory("Choose a Directory");
10  output = getDirectory("Choose a Directory");
11  suffix = getString("File suffix", ".tiff");
```

```
02b_Process_Folder_Dialog.ijm
```

Here `getDirectory` returns a string with the path pointing to the directory chosen by the user. `getString` returns a string entered by the user, or the default-string ".tiff" if nothing is changed by the user. The output strings are assigned to the variables `input`, `output` or `suffix`, respectively.

---

12 `02a_Process_Folder_Path.ijm`
13 `02b_Process_Folder_Dialog.ijm`

Note that there is also a family of macro commands that starts with `Dialog.`, which allow us to create more complex GUIs for user input. Please refer to the command reference[14] if you are interested in more details.

Script parameters

Another option is to use script parameters.[15]

```
9    #@ File (label = "Input directory", style = "directory") input
10   #@ File (label = "Output directory", style = "directory") output
11   #@ String (label = "File suffix", value = ".tiff") suffix
```

02c_Process_Folder_ScriptingParameters.ijm

Script parameters are by default used in the `Process_Folder.ijm` batch template — we had replaced them with the actual path in the macro with hard-coded path.[16]

`#@` initiate any Script Parameter, followed by the *Type* of variable. `#@ File`, for example, hands over a path to a file, `#@ String` a String etc. The appropriate dialog window is automatically generated according to *Type*. The dialog window can then be further customized using the options within a single parenthesis after `#@ Type`. As an example, see Line 11 of the code shown above. The options within parentheses specify the message within the dialog box (File suffix) and propose a default value ("".tiff""). You can find more information about script parameters on the ImageJ website.[17]

2.7.4 ImageJ Macro, Scijava

Once we use script parameters, we can take advantage of a very convenient way to execute code in batch.

Let us have a look on `03_SciJava.ijm`. When we compare it with the script parameter macro with loops,[18] we can see that both contain the same lines for choosing input- and output-folder, and suffix, via script parameters. `03_SciJava.ijm` also contains the same commands for saving the output files (Results window, binary image, ROI Manager), as discussed in ▶ Sect. 2.7.2. However, `03_SciJava.ijm` does not contain any code to batch execute the workflow — neither searching for files, nor looping over several files. To still be able to batch-execute this code, we click "batch" within the Script Editor. The GUI displayed in ◩ Fig. 2.3 opens. Select "Input" as parameter to batch and add files to the Input files list.[19] After clicking OK,

---

14 Built-in Macro Functions: ▶ https://imagej.nih.gov/ij/developer/macro/functions.html
15 02c_Process_Folder_ScriptingParameters.ijm
16 02a_Process_Folder_Path.ijm
17 ▶ https://imagej.net/Script_Parameters
18 02c_Process_Folder_ScriptingParameters.ijm
19 There are many ways to populate the list of files: (1) Select any number of files and drag them into the list field; (2) Use the "Add files…" button; (3) Select a folder and drag it onto the "Add folder content…" button; (4) Click the "Add folder content…" button and then select a folder.

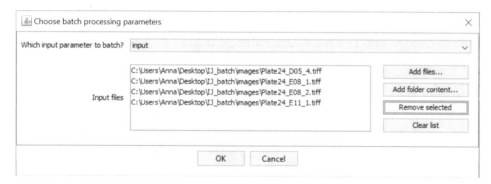

we see the dialog window for choosing the output directory and the file suffix. Once everything is set, all files listed in the "Input files" field will be processed.[20]

2.7.5 Command-Line Headless Methods

Why do we have script parameters? Macro commands include the `getString`, `getNumber` and `Dialog` family commands that allow us to create a user-interface – why should we need another way?

This is because script parameters are designed to be generic and universal in terms of interface. This means that the interface for input and output parameters can take any form, including GUI and Command-Line Interface (CLI). As we have seen already, script parameters only declare the type and the name of the variable. How these variables are provided is automatically determined depending on how the macro is called. If you run it with the "Run" button in the Script Editor, the input GUI is automatically generated and shows up on the screen. If you click the "Batch" button, the file names are automatically passed to the variable one-by-one, and images are processed in batch. This is in contrast to the `get` commands and `Dialog` commands of the Build-in macro functions, which are limited only to the input via dialog windows.

Now, using this generic and flexible characteristic of script parameters, we can try to run the batch processing macro from command line, without launching Fiji on your desktop. We call this way of running a software "the headless mode", as the main menu bar (here Fiji) never appears on your screen. This headless usage is especially important when you want to run the macro on a remote server, or a cluster without display.

---

20 Since the input is now handled via the Batch button, we do not need to specify label and style anymore as arguments of the Script Parameter. `#@ File (label = "Input directory", style = "directory")` input can therefore be shortened to `#@ File input`.

The command line interface is available in Windows (Git BASH), in Mac OSX (Terminal.app), and in Linux (e.g. Gnome).[21] The first thing to do is to create a command line alias for Fiji. This can be done with the following command:

Windows

```
alias fiji='/<Path-to-Fiji>/Fiji.app/ImageJ-win64.exe'
```

Mac OSX

```
alias fiji='/<Path-to-Fiji>/Fiji.app/Contents/MacOS/ImageJ-macosx'
```

Linux

```
alias fiji='/<Path-to-Fiji>/Fiji.app/ImageJ-linux64'
```

In all these cases, `<Path-to-Fiji>` should be replaced by the actual path to Fiji in your local machine. For example, if Fiji is located in the Applications folder of your Mac, the full path to the Fiji executable is:

```
alias fiji='/Applications/Fiji.app/Contents/MacOS/ImageJ-macosx'
```

Then, try the following:

```
    fiji --help
```

If your alias setting was successful, this command should print all the options that can be used in the CLI for running Fiji.

The second step is to prepare an example batch processing macro. Here, we can take an example from the Script Editor, just like we did in ▶ Sect. 2.7.2 *Macro IJ1*. Open a new Script Editor, and select the menu item [Template > ImageJ 1.x > Batch > Process Folder (IJ1 macro)]. Save the generated example as it is somewhere in your file system. Now, let us just run it from the Script Editor by clicking the "Run" button. You are asked for the locations of an input folder and an output folder, so please choose a folder that contains several TIFF images as the input directory, and choose any folder as the output. After clicking OK, you should see that all TIFF image file names appear printed in the Log window.[22]

Using exactly the same script that we tried above, let us run the macro in headless mode. Instead of setting the file paths to input and output folder in the dialog window, we can feed this information as options to the command.[23]

```
    fiji --ij2 --headless -run "<path-to-the-macro>"
    ↪  'input="<path-to-in-folder>",output="<path-to-out-folder>"'
```

Each option sets the following conditions:
- `--ij2` : use ImageJ2 instead of ImageJ1;
- `--headless` : run in headless mode;
- `--run <macro> [<arg>]` : run `<macro>` in ImageJ, optionally with arguments separated by comma.

---

21 In all these OS, command line interfaces by default use BASH, the most widely used Unix shell commands.

22 As this is a demo macro, it does not actually process and save images. It only shows that the macro can batch-access files in a folder.

23 For more details, see ▶ https://imagej.net/Scripting_Headless

Note how arguments of -run <macro> [<arg>] for paths to input and output folders are set in the above command. Both options are surrounded together by single quotes, and inside them, each path is surrounded by double quotes. The parts surrounded by single quotes are handed over as a single option to the script parameter resolver for macro. Before the execution of the macro, paths specified for variables input and output are separately interpreted as script parameters and used during the macro execution. In this way, options can be nested for different handling.

If you are successful in running the command, the CLI output will list the files from the selected input directory. The code can then be extended just like we have already done in previous sections to include the actual workflow.

Have on mind that macro that work on Desktop (GUI) sometimes fail to work in CLI. This is because some ImageJ functions are tightly associated with GUI and cannot run in the headless mode. For example, a macro that uses the ROI Manager does not work in the headless mode, as the ROI Manager relies heavily on GUI. In such a case, overlays can be used as an alternative to ROIs.

2.8 Collecting Measurement Results During Batch Processing

In this section, we discusses how to collect values that result from analysis of different images. As an example, we will collect the number of detected/analyzed nuclei per image and then calculate the mean and the standard deviation of the number of nuclei per image. We aim for the output of a style: "On average, there were 35.3 ± 10.5 nuclei analyzed per image (number of images=4).".

2.8.1 Collecting Measurements Within an Array

A popular way to collect the measurements is to use an array that is filled with values every time we execute our workflow while iterating over the images. Here, we refer to such an array as "data storage array". The data collecting macro[24] demonstrates the usage of such a data storage array. This macro is based on a macro that we already discussed, the one that uses script parameters and explicitly contains the code for looping.[25]

We start by creating an empty array in the beginning of the code.

```
17   collect_nNuclei = newArray();
```

04_CollectWithinArray.ijm

We do not specify the length of the array in this case, which is crucial since at this point of execution IJ has not searched for the files and does not know how many values we will collect.

To now fill the array step-by-step, we execute the following procedure:

---

24 04_CollectWithinArray.ijm
25 02c_Process_Folder_ScriptingParameters.ijm

1. We pass the data storage array `collect_nNuclei` as input parameter of the function `processFolder`. Additionally, we introduce the data storage array `collect_nNuclei` also as the returned value of `processFolder`. In this way, the data storage array is passed to `processFolder`, can be modified within the function and then the modified array `collect_nNuclei` will be returned as output. The final function call looks like this:

```
20  collect_nNuclei = processFolder(input, collect_nNuclei);
```

2. We extract the number of analyzed nuclei in each image. As discussed above, `processFolder` is searching for files (not directories) that end with a defined suffix − ".tiff". When such a file is found, the function `processFile` is called. `processFile` contains the image-processing workflow. We need to modify it in order to extract the number of analyzed nuclei. We get the number of analyzed nuclei by using `roiManager("count")`, and take this as the returned value of the `processFile` function using `return`:

```
95  nROIs = roiManager("count");
96  return nROIs; //output of the processFile function
```

Note again that we need to change how `processFile` is called:

```
36  nNuclei = processFile(input, output, list[i]);
```

3. In the final step we need to add the output `nNuclei` to our collecting array `collect_nNuclei`. This happens within the `processFolder` function: we extend the collecting array with the new value `nNuclei` by concatenation.

```
37  collect_nNuclei = Array.concat(collect_nNuclei , nNuclei);
```

❓ **Exercise 3**

Explain why we cannot create the data storage array within the `processFolder` function.

2.8.2 Collecting Measurements Within a Table

Very often we want to collect measurements from different images and save them in a Results Table. We can do so by creating a table, filling it up with the measured values, and once this is done, we can convert the table to an IJ1 Results table. We can easily add summary statistics to an IJ1 Results table using a native function.[26]

To do this, we first create a table, and initialize an index variable `rowIndex` for filling up the table.

```
18  Table.create("Numbers");
19  rowIndex = 0;
```

For adding a value to the table we use:

```
44  selectWindow("Numbers");
45  Table.set("nNuclei", rowIndex++, nNuclei);
```

---

26 04b_CollectWithinTable.ijm

Finally, in order to use the analysis tools available for a Results table we need to rename the table to "Results". With this, we can get the summary statistics of measured values:

```
26  selectWindow("Numbers");
27  Table.update;
28  Table.rename("Numbers", "Results");
29  run("Summarize");
```

2.8.3 Collecting Measurements When Using SciJava

In ▶ Sect. 2.7.4 we have seen how to utilize the Batch button to conveniently batch execute code using script parameters, without having to explicitly list the files and folders. Similarly, we can collect and output measurements in a very convenient way using script parameters.[27] Using the macro 03_SciJava.ijm as start, we only need to add an output parameter and assign our measurement of interest (number of nuclei) to the output variable. The output parameter is defined in the beginning using #@output

```
15  #@output nROIs
```

The number of nuclei is assigned to nROIs in the end of the workflow:

```
57  nROIs = roiManager("count");
```

For each image that we analyze, the output variable nROIs is added to a IJ2/SciJava table that is hidden during the macro execution. This is a different kind of table than the IJ1 Results table, and appears on the desktop when the macro is completed. The table can be saved as CSV file using the menu command [File/Export/Table...]. Just as any CSV file, it can be reopened in Fiji. If you rename the opened CSV file to "Results", you can also use the summarize-functions to calculate some statistics, as demonstrated in the previous section.

2.9 Application to Bioimage Analysis

▶ Example: Preparing Microscopy Image Files for Visual Inspection

When performing an imaging-based experiment, the first step of the analysis should be a visual inspection of the images. What do you see in the images taken under, or related to, different conditions (e.g. wildtype vs. mutant, treated vs. untreated)? Typical parameters in biology are e.g. changes in intensity of a protein of interest or changes in cell shape. For such an analysis, it is necessary to compare a few, ideally randomly chosen, images reflecting the different conditions side by side. In a standard microscopy experiment, that often means: opening the vendor-format via Bioformats in Fiji, selecting the same plane or channel in all image files, and setting the same brightness and contrast limits for all files. This is time-consuming and error-prone. However, all these steps can be easily recorded using the Command Recorder and then performed on all images in a folder. ◄

---

27 04c_CollectSciJava.ijm

We take as an example the 3-channel images already used in this chapter. For easy comparison we:

- Extract the signal channel (Channel 2) by duplication.
- Change the look-up-table (LUT) to gray, since a gray LUT is best inspected by a human eye.
- Set defined values as minimum and maximum contrast. This is essential for comparing the intensities for different images (see also Exercise 3).

These steps can be recorded:[28]

```
1  run("Next Slice [>]");
2  run("Duplicate...", "duplicate channels=2-2");
3  run("Grays");
4  setMinAndMax(0, 2000);
```

The only output we aim to save is the contrast-adjusted Channel 2. Therefore, the easiest solution would be to simply copy the code snippet from the Command Recorder to the Batch Processor and execute it after choosing input- and output-folder. The extracted and contrast-enhanced Channel 2 of the different image files are saved to the output folder. These visually enhanced images can then be re-opened in Fiji and easily visually compared. However, when we inspect the saved output files, we observe that they are saved under the same name as the original input files. This bears the risk of errors (e.g. deleting the original files by mistake).

In order to have more control over the saving process it is better to choose the Sci-Java solution discussed in ▶ Sect. 2.7.4. For this we need to specify the input and ask the user to select the output-folder via script parameters. Command `open(input)` opens the file. We then prepare for easy saving by extracting the filename without file ending. After the short workflow we then save the image and clean-up for the next image. This code runs in batch-mode when clicking on "Batch" within the Script Editor and after adding the files to batch process to the files list (see ◻ Fig. 2.3).

```
2   #@ File input
3   #@ File (label = "Output directory", style = "directory") output
4
5   open(input);
6   filename_pure = File.nameWithoutExtension;
7   saving_prefix = output + File.separator + filename_pure;
8
9   run("Duplicate...", "duplicate channels=2-2");
10  run("Grays"); //a gray LUT is best to inspect for the human eye
11  setMinAndMax(0, 2000); //defining fixed values for the image
    ↪   contrast.
12
13  saveAs("tiff", saving_prefix + "_C2.tif");
14
15  //Clean-up
16  run("Close All");
```

Example_Adapted_SciJava.ijm

---

28 Example_asRecorded.ijm

❓ **Exercise 4**

Which command is recorded when you click the "Auto" button in the Brightness/Contrast window? How does this command work and how does it compare to setMinAndMax? Why is it wrong to use "Auto" when we want to compare the intensity of a signal in different images?

┌─ **Take-Home Message** ─────────────────────────────────

In Fiji, there are various methods to construct batch-processing workflows. Each method has its own characteristics, advantages and disadvantages, and users can choose ones that best suit their needs in a given situation.

Solutions to the Exercises

✅ **Exercise 1**

```
title = getTitle();
```

Gets the name of the image and assigns it to the variable title.

```
run("Set Scale...", "distance=0 known=0 pixel=1 unit=pixel");
```

Removes the physical calibration, since it is incorrect (inches). All measurements are expressed in pixels.

```
Stack.setChannel(3);
```

Activates Channel 3 of the stack.

```
run("Duplicate...", "title=C3\_" + title);
```

Makes a copy of the activated Channel 3, naming it "C3_" + title.

```
run("Median...", "radius=10");
```

Smoothing of the duplicated Channel 3 by applying a median filter with radius 10. Median filtering preserves the edges of the nuclei.

```
setAutoThreshold("Li dark")
```

Calculation of a binary threshold using the auto-threshold method Li (Li and Tam, 1998). A most suitable auto-threshold method for the dataset was determined visually beforehand.

```
setOption("BlackBackground", true);
```

Command automatically recorded when using the "Threshold.." command in Fiji. Reflects the settings under [Process>Binary>Options].

```
run("Convert to Mask");
```

This line was automatically recorded when the "Threshold.." command was used in Fiji. Applies the automatically determined binary threshold and creates the binary mask.

```
run("Fill Holes")
```

Fills holes in the binary objects. Holes are background pixels fully surrounded by fore-ground pixels.

```
run("Set Measurements...", "area display redirect=None
↪   decimal=3");
```

Sets the type of measurements performed: We measure the **area**, **display** the label, mea-sure on the active image (`redirect=None`), and display the measured values with a precision of three digits below the decimal point (**decimals**).

```
run("Analyze Particles...", "size=1000-Infinity display exclude
↪   add");
```

Runs the [Analyze Particles...] command, which executes a connected com-ponent analysis. We **exclude** objects smaller than 1000 pixels, **display** the results in the Results Table, **exclude** objects that touch the border, and **add** the outline of the valid objects to the ROI Manager.

✅ Exercise 2

Folders are processed one after the other. Example of 3 layers of folders (paths shortened for clarity):

```
/Plate24_D05_4.tiff
/Plate24_E08_1.tiff
/Plate24_E08_2.tiff
/Plate24_E11_1.tiff
/subfolder1/Plate24_E08_2_level1.tiff
/subfolder1/Plate24_E11_1_level1.tiff
/subfolder1/subfolder2/Plate24_D05_4_level2.tiff
/subfolder1/subfolder2/Plate24_E08_1_level2.tiff
```

✅ Exercise 3

Inside the function `processFolder`, we call `processFolder` recursively when the path in list[i] is a directory. This allows the processing of all files in all subfolders. Cre-ating collect_nNuclei by `collect_nNuclei = newArray()` within the function `processFolder` would cause overwriting of `collect_nNuclei` each time when a new subfolder is processed.

✅ Exercise 4

When clicking "Auto" in the Brightness/Contrast window, the following command is recorded: `run("Enhance Contrast", "saturated=0.35")`. The second argu-ments indicates that the saturation of pixel values is 0.35, which means that the 0.35% darkest and brightest pixels of the image will all be set to 0 and 65535 (in case of 16-bit image), respectively, by computing appropriate minimum and maximum pixel values to satisfy the requested percentage of pixels to become saturated. At the same time, other pixels with values between these minimum and maximum become scaled linearly. This means that images with different brightness will be scaled differently e.g. a darker image will be enhanced more. Consequently, by applying "auto-contrast", an image with the maximum value = 500, would look similar to an image with the maximum value of 1500 by different degree of enhancements. Thus, applying auto-contrast is quite misleading if one needs to compare images to inspect the difference in the intensity of a structure, e.g. the expression level of a protein.

To scale and enhance images to the same degree, we can specify the minimum and the maximum values by using `setMinAndMax(0, 2000)`. With this command, we are scaling all images using fixed limits (minimum and maximum pixel values) and by that we can compare the images after the enhancement. Note that these limits should fit within the range of all pixel values in all images.

Acknowledgements Images were recorded with the help of Susanne Hasse, Mihail Sarov lab, MPI-CBG, Dresden, Germany. We thank Jan Eglinger (FMI Basel, Switzerland) for thoroughly reading the text, testing the code, and giving valuable suggestions for further improvements.

Further Readings The textbook "Bioimage Data Analysis", (Miura et al., 2016), contains a chapter aimed at helping to learn ImageJ macro language. If you are not familiar with this language, please consider going through that chapter. The book is freely downloadable from the website: ▶ https://bit.ly/bias-wiley

References

Li CH, Tam PKS (1998) An iterative algorithm for minimum cross entropy thresholding. Pattern Recognit Lett 19(8):771–776

Miura K, Tosi S, Möhl C, Zhang C, Paul-Gilloteaux P, Schultz U, Nørrelykke SF, Tischer C, Pengo, T (2016) Bioimage Data Analysis. Wiley-VCH, Weinheim. ▶ https://analyticalscience.wiley.com/do/10.1002/was.00050003/full/

Rueden CT, Schindelin J, Hiner MC, DeZonia BE, Walter AE, Arena ET, Eliceiri KW (2017) ImageJ2: ImageJ for the next generation of scientific image data. BMC Bioinform 18(1):529. https://doi.org/10.1186/s12859-017-1934-z

Schindelin J, Arganda-Carreras I, Frise E, Kaynig V, Longair M, Pietzsch T, Preibisch S, Rueden C, Saalfeld S, Schmid B, Tinevez J-Y, White DJ, Hartenstein V, Eliceiri K, Tomancak P, Cardona A (2012) Fiji: an open-source platform for biological-image analysis. Nat Methods 9(7):676–682. Publisher: Nature Publishing Group, a division of Macmillan Publishers Limited. All Rights Reserved. https://doi.org/10.1038/nmeth.2019

Schneider CA, Rasband WS, Eliceiri KW (2012) NIH Image to ImageJ: 25 years of image analysis. Nat Methods 9(7):671–675. Publisher: Nature Publishing Group ISBN: 1548-7091. http://www.nature.com/doifinder/10.1038/nmeth.2089

Python: Data Handling, Analysis and Plotting

Arianne Bercowsky Rama

Contents

This Chapter has been reviewed by MartinWeigert, École Polytechnique Fédérale de Lausanne, and Uwe Schmidt, Myers lab, Center for Systems Biology Max Planck Institute for Molecular Cell Biology and Genetics, Dresden.

K. Miura, N. Sladoje (eds.), *Bioimage Data Analysis Workflows–Advanced Components and Methods*, Learning Materials in Biosciences, https://doi.org/10.1007/978-3-030-76394-7_3

3

What You Will Learn in This Chapter

When performing an image analysis pipeline, a programming language like Python is mainly used for two distinctive applications: (1) the analysis of the acquired images, such as background removal, noise reduction, object segmentation, measurements of biological structures and events, etc. and (2) the analysis of the data obtained as a result of the image analysis, such as a calculating a histogram from the noise-removed image or statistics on the shape of the segmented object. The aim of this chapter is to show how Python can be used as a tool to analyze the data obtained as the final step of a bioimage analysis workflow. We will learn how to arrange the data into a tidy form, which is a way to structure the data to simplify the later analysis. Python libraries `pandas`, for data handling, and `bokeh` and `holoviews`, for data plotting, are discussed along this chapter. Jupyter notebooks are fully available to follow the examples, however, minimal Python knowledge is required (the concepts of Python lists, dictionaries and arrays should be known).[1]

3.1 Tools to Follow the Chapter

This chapter uses Anaconda[2] as the Python Distribution and Jupyter Notebooks[3] to run the Python code. As mentioned, Jupyter Notebooks (specified at the beginning of each section) are available for the reader to follow the examples and to try out the Python code:

1. NB-0-Installation_Guide.ipynb: Installation of Python distribution and all the packages needed to follow the chapter.
2. NB-1-Python_Introduction.ipynb: Brief introduction to basic operations in Python, which will be useful if you are new to Python.
3. NB-2-Pandas_Data_Handling.ipynb: This notebook covers ▶ Sect. 3.3, how to handle data using the package `pandas`.
4. NB-3-Bokeh_Plotting.ipynb: This notebook covers ▶ Sect. 3.4, specifically 3.4.1—using `Bokeh` to create interactive figures.
5. NB-4-Holoviews_Plotting.ipynb: This notebook covers ▶ Sect. 3.4, specifically 3.4.1—using `HoloViews` to create interactive figures.

These notebooks are available in Github.[4]

---

1   This chapter was communicated by Uwe Schmidt (Center for Systems Biology MPI-CBG, Dresden, Germany) and Martin Weigert (EPFL, Lausanne, Switzerland).

2   ▶ https://www.anaconda.com/products/individual.

3   ▶ https://jupyter.org.

4   ▶ https://github.com/NEUBIAS/neubias-springer-book-2021/tree/master/Ch03_Python_Data_handling_analysis_and_plotting.

3.2 Why Python?

Python is a high-level programming language designed in the early 1990s by Guido van Rossum. It executes instructions without the need of a compiler—i.e., it is an interpreted language—and its operations are done at run-time—i.e., it has dynamic semantics—making it a fast language to prototype in. For at least two decades it has been widely used, which makes it beginner-friendly due to the amount of tutorials and documentation that exist on the web. In fact, in the past years Python has shown a huge growth in demand, due to the increase in:

- Publications: Books, conferences, journals;
- Users: Number of downloads and number of uses. A trend calculated by Stack-Overflow,[5] which counts the tags and posts on the platform, shows a very high and still increasing popularity of the Python language.
- Applications: web and internet development, scientific and numeric computing, education (teaching programming), desktop GUIs (graphical user interface), software development among other applications.[6]

The purpose of this chapter is to show one way to use Python as a tool to analyze data and obtain browser-interactive figures which are easy to share. We will use Anaconda as our Python distribution to simplify the package installations. As our web-based application for writing and running Python code we will use Jupyter Notebooks, which combine code with narrative text, equations, and visualizations. There are some other great notebook alternatives, like for example Google Colaboratory[7] (Colab for short) that allows the execution of Python in a browser without the need for any prior installations and with free access to GPU.

3.2.1 Python Versions

Since the first release in 1994, there have been several Python versions. Newer versions add features either in the language itself, its built-in functions, or in standard library support modules (Mertz, 2015). The two most recent versions are Python 2 and 3. Python 2 has not received further updates or bug-fixes as of January 2020.[8] In this chapter we will be using Python 3.6 or higher.

3.2.2 Python Packages and Environments

Python, like many other programming languages, allows modular programming. This means that the code can be broken down to create smaller and more manageable scripts named modules. Grouping these modules can then result in a Python package. For example, NumPy (▶ https://numpy.org) is a package for scientific computing

---

5 ▶ https://stackoverflow.com.

6 ▶ https://www.python.org/about/apps/.

7 ▶ https://colab.research.google.com.

8 ▶ https://www.python.org/doc/sunset-python-2/.

(Harris et al., 2020) which we will be using later in this chapter. Depending on what we want to achieve, we will need different packages that already exist.

There are several ways to install packages which will be explained later. Once the packages are installed, in order to use them we have to make them available in our code. For example, to use all the functions in the NumPy package, we first need to import NumPy using the import statement:

```
[1]: import numpy as np
```

Here we have imported the package NumPy which is now bound to the name we have chosen, np (which in this case is standardized). This means that whenever we want to call a NumPy function, e.g. to calculate the square root of 4, we will use:

```
[2]: np.sqrt(4)
```

The same way Python has different versions, the packages have them as well. Depending on the project we work on, we might need different package versions. However, it could be problematic if two different projects need different versions. This is where the *environments*[9] are very useful. They allow the creation of an isolated environment for each of the different projects, where the package versions are independent in each environment. There are several ways to set up a virtual environment depending on the tools used to run Python. Later in this chapter we will learn one of the many ways to do so.

3.2.3 Anaconda

Jupyter Notebook: `NB-0-Installation_Guide.ipynb`

Python and installation of packages can sometimes be complicated, which is why here we describe an easy way to do so, with the minimal amount of potential problems. There are many ways to set up a Python environment for scientific computing or for any other purpose. Two common ones are:

1. Installing packages on demand from the Python Package Index (PyPI), a repository of software for the Python programming language. As of today, there are more than 250,000 packages which can be downloaded from PyPI using the package installer `pip`.
2. Downloading a Python distribution that already contains many of the most popular packages needed. One of the major distributions, and the one we are using in this chapter, is Anaconda[10] which contains `conda` to manage and install packages. You could also install Miniconda,[11] which is a free minimal installer using `conda`.

---

9   ▶ https://docs.python.org/3/library/venv.

10 ▶ https://www.anaconda.com.

11 ▶ https://docs.conda.io/en/latest/miniconda.

3

> ☐ **Table 3.1** Main differences between `pip` and `conda`. For a more detailed explanation, visit
> ▶ https://www.anaconda.com/blog/understanding-conda-and-pip

`pip`	`conda`
Installs packages from PyPI	Installs packages from Anaconda Repository and Anaconda Cloud
Installs Python Packages	Installs packages written in any language
Python interpreter must be installed before using pip	Installs Python packages as well as the Python interpreter directly
Has no built-in support for environments—relies on tools like `virtualenv` or `venv`	Is also an environment manager

`pip` is the recommended tool for installing packages from PyPI. `pip` installs Python software, but may require that the system has compatible compilers, and possibly libraries, installed before invoking `pip`. Another installer tool is `conda`, which can handle both Python and non-Python installation tasks. `conda` is an open-source cross-platform package and environment manager that can install and manage packages from the Anaconda repository, Anaconda Cloud and other channels such as conda-forge.[12] There is never a need to have compilers available to install `conda` packages. Additionally, as mentioned before, the packages may also contain C or C++ libraries, R packages, or any other software.

We will use `conda` to install the packages for this chapter. However, it is good to understand the main differences between these two package managers—`pip` and `conda`—to know when to use which of them. They are summarized in ☐ Table 3.1.

As mentioned earlier, we will use Anaconda (▶ https://docs.anaconda.com/anaconda) as our Python distribution to simplify Python and package installations. Moreoever, Anaconda is a package manager, an environment manager, a Python/R data science distribution, and a collection of over 7500 open-source packages. It was created with the aim to simplify package management and deployment. Package versions in Anaconda are managed by the package management system `conda`. It also includes a graphical user interface (GUI), Anaconda Navigator (☐ Fig. 3.1), which is an alternative to the command-line interface.

3.2.4 Jupyter Notebook

Once Python is installed, there are many ways to run Python code, for example using the command-line or terminal by typing in `python` (or `python3`, depending on the installation) and hitting enter. However, in this chapter we will run Python code in a web-browser in a way which allows that we mix code, text, and equations, such that it resembles a notebook.

When Anaconda is installed, we get Python installed, and—conveniently—in addition we get installed several commonly used packages for scientific computing and

---

12 ▶ https://conda-forge.org.

Anaconda GUI

Jupyter Notebook

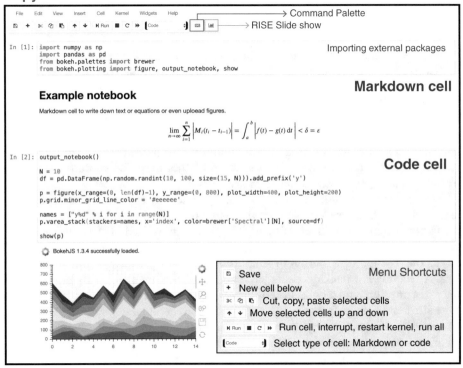

◻ Fig. 3.1 Upper panel: Anaconda GUI included with the Anaconda distribution. It contains, among others, Jupyter Lab (which is a more interactive version of the Jupyter Notebook), Jupyter Notebook (which we will be using in this chapter) and Spyder which is more similar to Matlab (it contains a variable explorer which resembles Matlab work-space). In the Anaconda Navigator we can manage the environments and the packages. We can also do this by using the Anaconda Prompt (command line shell). **Lower panel**: Example of a Jupyter notebook with the two main types of cell: Markdown, for text and equations and Code, for writing Python code (you could also set it up to write R, Julia, Groovy, Java...). The Command Palette shows keyboard shortcuts

data science and some applications, including Jupyter Notebook (which can also be installed without Anaconda, using `pip`).

Project Jupyter is a non-profit, open-source project, born out of the IPython Project (▶ https://ipython.org) in 2014, as it evolved to support interactive data science and scientific computing across many programming languages (▶ https://jupyter.org). Jupyter Notebooks allow to write code, Markdown text and equations and save the notebooks as Hypertext Markup Language (HTML) or even as Portable Document Format (PDF). Figure ◻ 3.1 shows an example of a Jupyter Notebook and some of the basic commands to start using it. However if this is the first time you are using Jupyter Notebook, you might want to check the Project Jupyter recommended documentation: ▶ https://jupyter.readthedocs.io.

Once we have Anaconda Distribution and we have downloaded all the packages and ran a Jupyter Notebook, we are ready to start handling and plotting data in the following sections. If you have not done this yet, NB-0-Installation_Guide.ipynb will guide you through the installation steps.

3.3 pandas: Python Data Analysis Library

Jupyter Notebook: NB-2-Pandas_Data_Handling.ipynb

As part of an image analysis pipeline, we will likely be handling and analyzing measurements of experimental image data. One of the most time-consuming parts is often arranging the data so that it is in a suitable format to perform the analysis and visualization of the results. pandas is a powerful tool for working with tabular data in the Python ecosystem. This section describes the use of pandas and how to arrange the data in a *tidy* format to make the analysis and visualization easier.

pandas is an open source library which allows efficient manipulation, reading and writing of (tabular) data. It was initiated by McKinney et al. (2011) and since then, it has been widely used in the Python community with the aim to be a *fundamental high-level building block for doing practical, real world data analysis in Python* (▶ https://pandas.pydata.org/). pandas makes it easy to work with labeled data: we can handle and arrange the data but, we can also label information on the data points, making it a powerful tool for handling *metadata*.[13]

The standard way to import pandas package is by using:

```
[1]:   import pandas as pd
       import numpy as np
```

Moreover we will also use the NumPy package which is why we are also importing it at the beginning of our code. In the following sections we will explore the power of pandas primary data structure, the DataFrame. We will also learn how to import/export data with pandas and how to arrange the data so that it is easier to perform statistical analysis and plotting.

---

13 Data that provides information about other data, e.g. the metadata for a microscopy movie could be the pixel size, image dimensions, acquisition settings like laser power, exposure time, etc.

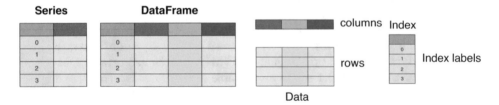

◻ Fig. 3.2 Basic structure of a DataFrame and a Series. The name of each component is important—we will be using them along the chapter

3.3.1 Syntax: Creating a DataFrame

pandas library is built on top of the NumPy package, which means that most of the NumPy functions are available for the pandas objects. However, what makes pandas so useful with respect to NumPy objects is the way the data is structured. pandas data structures have rows and columns with a similar appearance as the tables in Excel or CSV files (among others), which makes statistical analysis easier. But before we get into more complicated data wrangling methods, we first define the most fundamental units of the pandas data structures: a Series and a DataFrame.

Series

pandas has two main data structures: Series, for 1-dimensional labeled data, and DataFrame, for 2-dimensional labeled data. They have similar structure: index column, column(s) and rows (◻ Fig. 3.2). Each column has a name associated with it, also known as label.

A Series is the simplest concept, therefore we will start by understanding how we can create one. The following line of code shows how to initialize a Series.

```
[2]:  pd.Series(data,index,dtype)
```

Here, data can be a Python dictionary, a NumPy array or a scalar value. The next parameter, index, is a list of axis labels (which is not the same as the column label). If no index is passed, one will be created having values [0, ..., len(data) - 1]. Also, as a NumPy array, a pandas Series supports dtype which can be float, int, bool, etc.

Here are three examples of how to initialize a Series:

```
[3]:  pd.Series(np.random.randn(7), index=["N", "E", "U",
        "B", "I", "A", "S"])
```

```
[3]:  N    -0.995606
      E    -1.160779
      U    -0.454513
      B    -0.590617
      I    -0.699399
```

```
A      2.248658
S     -0.189257
dtype: float64
```

```
[4]:  pd.Series({"A":0, "B":1, "C":2}, dtype=float)
```

```
[4]:  A      0.0
      B      1.0
      C      2.0
      dtype: float64
```

```
[5]:  pd.Series(10, index=["a", "b", "c"], dtype=int)
```

```
[5]:  a      10
      b      10
      c      10
      dtype: int64
```

A `Series` is a NumPy array-like, which means that it can be passed into most NumPy methods expecting a NumPy array. However, a key difference between pandas `Series` and NumPy ndarray is that operations among `Series` automatically order the data based on the index. Therefore, if we need an actual ndarray, we can use the command `Series.to_numpy()`.

DataFrame

The most commonly used `pandas` concept is a `DataFrame`, a 2-dimensional labeled data structure with columns of potentially different types. Similar to a `Series`, a `DataFrame` object can be created using the following line of code.

```
[6]:  pd.DataFrame(data,index,columns,dtype)
```

The `data` can be a Python Dictionary of 1D arrays, lists, dicts or `Series`, as well as a 2-D NumPy array or another `DataFrame`. The `DataFrame` has labeled axes: `rows` (axis=0) and `columns` (axis=1). The rows and columns can be accessed by the index and columns attributes, respectively: `DataFrame.index` and `DataFrame.columns`

Once the `DataFrame` has been defined, we can select, add, and delete columns in similar ways as a Python dictionary.

```
[7]:  df = pd.DataFrame({"A":["a", "b","c"], "B":[1,2,3]})
      df
```

```
[7]:       A   B
      0    a   1
      1    b   2
      2    c   3
```

3

```
[8]:  # Add a column
      df["C"] = ["D","F","G"]
      df
```

```
[8]:     A  B  C
      0  a  1  D
      1  b  2  F
      2  c  3  G
```

```
[9]:  # Delete a column
      del df["A"]
      df
```

```
[9]:     B  C
      0  1  D
      1  2  F
      2  3  G
```

By default, a column is inserted at the end of the DataFrame. However, using the insert function, we can specify the location (loc) of the new column and the values we want to insert.

```
[10]:  df.insert(loc,column,value)
```

DataFrames are indexed by columns, df[column_name], but we can also select both rows and columns by using:

```
[11]:  df.insert(row_index,column_name)
```

There are several ways to index a DataFrame; some of them are summarised in ▣ Table 3.2.

3.3.2 Basic Numeric Operations

Pandas has methods and functions to carry out binary operations[14] for matching and broadcasting behaviour. In the following example we initialize two DataFrames, df1 and df2, using two dictionaries, d1 and d2:

```
[12]:  # Dictionary
       d1 = {"abc": ["a", "b", "c", "a", "b", "c"], \
             "123": [1, 2, 3, 4, 5, 6], \
             "ABC":["A", "A", "B", "B", "C", "C"], \
             "num": ["one", "two", "three", "four",
             "five", "six"]}
```

---

14 Binary operation: calculation that combines two elements to produce another element.

> ▣ **Table 3.2** Indexing a `DataFrame` is intuitive to help getting and setting subsets of the data-set. For more information on indexing `DataFrames`, visit ▶ https://pandas.pydata.org/pandas-docs/stable/user_guide/indexing.html

Syntax	Description
`df[column name]`	Select a column. Results in a `Series`.
`df[[column names]]`	Select one or more columns. Results in a `DataFrame`.
`df.loc[label]`	Select a row by a label. Results in a `Series`.
`df.iloc[loc]`	Select a row by an integer location. Results in a `Series`.
`df[2:5]`	Slice the rows. Results in a `DataFrame`.

```
d2 = {"abc": ["d", "d", "e", "c", "b", "e"], \
      "123": [7, 8, 9, 10, 11, 12], \
      "ABC":["D", "D", "E", "A", "B", "A"], \
      "num": ["seven", "eight", "nine", "four",
      "five", "six"]}

# DataFrame from a Dictionary
df1 = pd.DataFrame(d1)
df2 = pd.DataFrame(d2)

df1
```

```
[12]:    abc  123 ABC    num
      0    a    1   A    one
      1    b    2   A    two
      2    c    3   B  three
      3    a    4   B   four
      4    b    5   C   five
      5    c    6   C    six
```

Some basic binary operations are addition `add()`, subtraction `sub()`, multiplication `mul()`, and division `div()`. The following example shows the addition of a column of `df1` and a column of `df2`.

```
[13]: df1["123"].add(df2["123"])
```

```
[13]: 0     8
      1    10
      2    12
      3    14
      4    16
```

```
5      18
Name: 123, dtype: int64
```

The same result can be achieved by computing df1["123"]+df2["123"], but using the add() method allows us to choose the dimensions and labels we want to use. The axis parameter allows using index (axis=0) or columns (axis=1) for the addition operation. Moreover, in the case of missing data, there are operations that include the parameter fill_value. If fill_value=0, the missing values, which in DataFrame by default are NaN values, are treated as zeros. If the same values are missing in both DataFrames, they will continue to be NaN. When computing df1["123"]+df2["123"], if there is a missing value, a NaN will be added in that position.

With a Series or a DataFrame it is very simple to compute descriptive statistics, e.g., as the mean value is computed in the following line of code:

```
[14]:  df.mean(axis,skipna,numeric_only)
```

In this example we apply the mean operation to the axis we choose (axis = 0 for index, axis = 1 for columns). In case there are missing values which have been replaced with a NaN, the skipna parameter, which is true by default, will exclude all NaN values from the computation. Finally, we can choose whether we want to include only float, int or boolean columns in the calculation, by specifying the parameter numeric_only. There are many other descriptive statistics; some examples are shown in ◘ Table 3.3. For more examples, visit the website.[15]

◘ **Table 3.3**   Examples of descriptive statistics for DataFrame and Series

Function	Description
count	Number of non-NaN observations
sum	Sum of all values
mean	Mean of all values
median	Median of all values
std	Standard deviation of all values
min	Minimal value
max	Maximal value
describe	Generates descriptive statistics
T	Transpose index and columns

---

15 ► https://pandas.pydata.org/pandas-docs/stable/user_guide/basics.html#basics-stats.

3.3.3 **Import Data Using pandas**

In the previous section we learned how data is structured in the pandas Series and DataFrame. When performing image analysis tasks we will most likely be using some other software, such as Fiji (Schindelin et al., 2012), to perform segmentation, cell tracking, protein co-localization analysis, etc. The outcome of such analysis usually comes in the form of a table. A usual next step is to export this table-like data into some software, such as R, Python, MATLAB, etc, to extract useful information.

With pandas we are able to read and write different data types: Microsoft Excel files (pd.read_excel(), pd.to_excel()), comma separated values files—CSV (pd.read_csv(), pd.to_csv()), JSON files (pd.read_json(), pd.to_json()), HTML (pd.read_html(), pd.to_html()), HDF5 (pd.read_hdf(), pd.to_hdf()), and more.

In this chapter, we focus on CSV files, since they are easy to read into data structures in many programming languages. As a general rule, we should always try to save the data in file formats that are open and readable in many contexts regardless of the specific software of choice.

To read a csv file into a DataFrame, we use the following line of code:

```
[15]:  pd.read_csv(filepath_or_buffer, sep, usecols,
        manage_dupe_cols, na_values)
```

Here we show only some of the many parameters to choose from the CSV reader. They help creating a DataFrame that best describes the data. To check all of the available parameters, visit the website.[16]

- filepath_or_buffer: Any valid string path.
- sep: Delimiter to use. By default, it is assumed that the data is separated by commas (sep=",").
- header: Row number(s) to use as column names for the DataFrame. For example, (header=[0,1,3]) will use the rows 0, 1 and 3 as headers, and will skip row 2. The default is to use the first row as column header (header=0).
- usecols: Returns a subset of the columns. For example, using integer indices of the data columns usecols=[0,1,2] or strings that correspond to the names of the columns in the data ["A", "B", "C"].
- mangle_dupe_cols: If there are two or more columns with the same name, by default they will be written as "Col", "Col.1", "Col.2". If mangle_dupe _cols=False, columns with the same name will be overwritten.
- na_values: Additional string values to be recognized as NaN. By default, any blank space will be recognized as a NaN, but also some other strings such as <NaN> and nan. This allows to apply statistics in a missing-value-friendly manner. This option allows other strings to be specified to also be included in the DataFrame as a NaN.

Once the data is imported and we are satisfied with the DataFrame we created, the next step, that helps to get the most out of the data, is to "tidy" this data-set. In the following section we will learn how to accomplish this with our already created, or imported, DataFrame.

---

16 ▶ https://pandas.pydata.org/pandas-docs/stable/reference/api/pandas.read_csv.html.

3.3.4 Reshape the Data: How to Create Tidy Data

Great part of the time and effort invested in analyzing a data-set goes to organizing the data and handling missing values, among many other preparation steps performed every time new data is collected. The way we build, e.g., an Excel file, is the most intuitive way for human perception, however, we should always try to convert the data into their tidy form. *Tidy data-sets* are data-sets that are arranged such that each variable is a column and each observation is a row (Wickham et al., 2014). This section gives a general explanation of what is tidy data, how it can be accomplished, and some of the benefits of analyzing a tidy data-set. Wickham defines data tidying as a standard way to clean the data. This allows us to map the meaning of a data-set to its structure (its physical layout). This structuring of the data facilitates the analysis specially if one is using vectorized programming languages, such as R or Python with NumPy. Specifically, tidy data complements panda's vectorized operations. We will see some examples in the following sections.

A messy data-set would be any other arrangement of the data. In ◻ Table 3.4, we have two examples of typical representations of messy data-sets. The data-table on the left represents results of a titration experiment in which the goal is to check how a measurable, e.g., mean fluorescence intensity of a gene expression marker, changes with different pulse duration (columns) and drug concentrations (rows). In this table, both the columns and the rows are labeled. The data-table on the right represents a similar experiment in which the same measurable should be checked, but in this case using two concentrations of DMSO (Dimethyl sulfoxide) as control and two concentrations of a drug being tested. In this case we observe what is called *multi-index*, with two levels of columns.

To convert the examples of messy data shown in ◻ Table 3.4 into their tidy forms, we need to identify the variables which should form the columns in our tidy data-set. In the first case, the pulse duration and the concentration of the treatments will be the two variables (the measures of two attributes). In the second case we will have two different treatments: Drug and DMSO and the concentration of these treatments: 0.1%, 0.5%, $10\,\mu$M, $50\,\mu$M. Following this rearrangement, we can obtain a corresponding tidy data-set (◻ Table 3.5):

Now we know what a tidy data-set is. The next step is to learn how to implement pandas functions to transform the structure of the data into a cleaned and ready-to-analyze tidy form.

◻ **Table 3.4** Examples of two messy data sets. Table to the left includes labeled rows and columns. Table to the right contains multi-index: two concentrations for DMSO treatment and two more for the Drug treatment

Concentration	5 min	10 min	20 min	30 min
500 μM	2.3	9.2	12.5	16.9
100 μM	5.4	9.9	13.3	17.0
20 μM	3.2	9.8	13.5	17.4
10 μM	4.8	9.2	14.2	17.7

DMSO		Drug	
0.1 %	0.5 %	10 μM	50 μM
20	100	32	78
28	102	47	98
34	103	53	96

◻ **Table 3.5** Example of a tidy form of the data-sets. In both cases (left and right), the columns in the tables are variables, whereas rows are observations: the result of one pulse duration with a specific drug concentration (left), and the result of a concentration from a given treatment (right)

Concentration	Pulse Duration	Result		Treatment	Concentration	Result
500 μM	5 min	2.3		DMSO	0.1%	20
100 μM	5 min	5.4		DMSO	0.1%	28
20 μM	5 min	3.2		DMSO	0.1%	34
10 μM	5 min	4.8		DMSO	0.5%	100
500 μM	10 min	9.2		DMSO	0.5%	102
⋮	⋮	⋮		⋮	⋮	⋮
10 μM	30 min	17.7		Drug	50 μM	96

Changing the Layout of the Data-Set to Get Tidy Data with `pandas`

One of the most useful functions to tidy our data-sets is `pd.melt(df)`. This function allows us to gather columns into rows from a `DataFrame`, which means to go from wide format (like in table ◻ 3.4) to long format (like in table ◻ 3.5). One thing to consider before melting the `DataFrame` is to specify what are the values and what are the variables:

```
[16]: pd.melt(DataFrame, id_vars, value_vars, var_name,
      value_name, col_level)
```

◻ Figure 3.3 illustrates the meaning of each of these parameters and how they will help to reshape our data into a tidy form.

The data in its tidy form is convenient for analysis. However, once we have finished all the analysis, we might want to have the data back in a form which is prettier to visualize as a table. To go in the opposite direction, i.e., from long to wide format, we can pivot our `DataFrame`:

```
[17]: pd.pivot(DataFrame, index, columns, values)
```

◻ Figure 3.3 also shows the parameters from the `pd.pivot(df)` function, used to reshape the data into a wide form.

Going back to the messy data examples from the previous section (◻ Table 3.4), we can use `pandas` function `pd.melt()` to convert them into their tidy form (◻ Table 3.5).

```
[18]: df1 =
      pd.read_csv("./Data/PulseVsConcentration.csv")
      df1_melted = pd.melt(df1, id_vars="Concentration",
      var_name="Pulse Duration", value_name="Result")
```

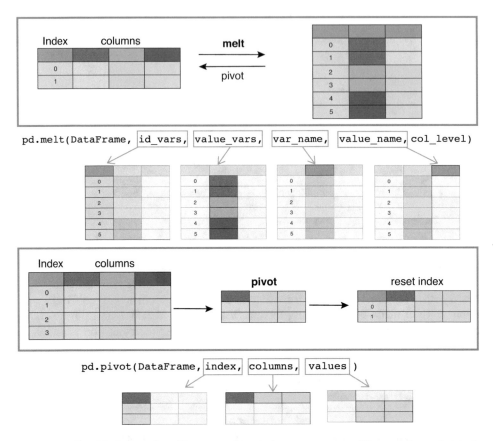

◻ Fig. 3.3 Graphical examples of how to melt and pivot DataFrames. Here we show what each parameter represents for the methods melt and pivot, to better understand how the data can be re-arranged and re-shaped. Inspired from the cheat-sheet by Irv Lustig from Python Data Wrangling Cheat-sheet (▶ https://pandas.pydata.org/Pandas_Cheat_Sheet.pdf)

After reading and saving the CSV file data into a DataFrame, we use the pd.melt() function. "Concentration" is already a column, so we assign it as our identifier variable. Next, we create a variable column called "Pulse Duration", which is currently the first row (5 min, 10 min, 20 min and 30 min). Finally, we rename the last column, which contains the intensity measured values, as "Result".

```
[19]: df2 = pd.read_csv("./Data/DMSOVsDrug.csv",
      header=[0,1])
      df2_melt = pd.melt(df2, var_name=["Treatment",
      "Concentration"], value_name="Result")
```

In this second example we also read and save the CSV file data into a `DataFrame`, but in this case, we specify that the first two rows contain the variables which will become the headers. The next step is to melt the `DataFrames`. In this case there are two rows which we have to convert into variable columns: "Treatment" (DMSO and Drug) and "Concentration" (0.1%, 0.5%, $10\mu M$, $50\mu M$). Finally, we rename the last column "Result" which contains the measured intensity for a treatment at a given concentration.

Now we have created tidy data, and we can manipulate, model, and visualize it easily and effectively. In the following section we will learn how to manipulate a tidy data-set.

3.3.5 Split-Apply-Combine

Usually, we perform some analysis based on some attributes of the data that we want to compare, or extract meaningful information, by performing statistical and/or numerical analysis. Intuitively, to do so, we (1) split the data into groups according to some criterion; (2) apply some functions to analyze the split-up data; and, finally, (3) combine the results to be saved in a new data set. The good news is that there is a conceptual framework to apply these steps—it is called the Split-Apply-Combine strategy, and was first formalized by Wickham et al. (2011). In this article Wickham describes the strategy as: "break up a big problem into manageable pieces, operate on each piece independently, and then put all the pieces back together". An R package was created with this strategy, but now `pandas` has its own way to implement the same idea.

This strategy only makes sense if the data is in a tidy format, because it will be split-up according to the selected columns. Therefore, we can apply functions to this newly grouped data and combine the results into a new data-set.

For an extensive tutorial on how to apply the split-apply-combine strategy using `pandas`, please visit the website.[17]

Split

The `df.groupby` operation performs the splitting step using any data axis. It allows the grouping of a `DataFrame` usually by one or more of the columns. The result is a `DataFrameGroupBy` object. Let us take the two `DataFrames` from ◻ Table 3.5 (in a tidy format) as an example; we can apply simple grouping operations:

```
[20]:   # (1) Group the melted DataFrames according to some
        # category (column)
        df1_groupby =
        df1_melt.groupby("Pulse Duration")
        # Group df2 by the Treatment category
        df2_groupby = df2_melt.groupby("Treatment")
```

---

17 ▶ https://pandas.pydata.org/pandas-docs/stable/user_guide/groupby.html.

```
[21]:  # Show the groups from groupby method
       print("df1 groups: ", df1_groupby.groups.keys())
       print("df2 groups: ", df2_groupby.groups.keys())
```

```
[21]:  df1 groups:   dict_keys(['10 min', '20 min', '30 min',
       '5 min'])
       df2 groups:   dict_keys(['DMSO', 'Drug'])
```

Once the data is grouped, we can split it by using the df.get_group() method. For example, we grouped the melted DataFrame df1_melt according to pulse duration, which gave rise to the groups: 5 min, 10 min, 20 min and 30 min. Now, we can get one of these groups.

```
[22]:  df1_groupby.get_group("10 min")
```

```
[22]:      Concentration  Pulse Duration   Result
       4          500 um          10 min      9.2
       5          100 um          10 min      9.9
       6           20 um          10 min      9.8
       7           10 um          10 min      9.2
```

Apply

Once the data has been grouped and split-up, we can apply different functions to the newly created DataFrames. For this, we may use one of the following operations:

1. Aggregation: Computes one or more summary statistics to the group_by object. The following example computes the mean and the sum using a NumPy function.

```
[23]:  df1_groupby.agg([np.mean, np.sum]).reset_index()
```

```
[23]:    Pulse Duration   Result
                            mean     sum
       0         10 min    9.525    38.1
       1         20 min   13.375    53.5
       2         30 min   17.250    69.0
       3          5 min    3.925    15.7
```

By default, the grouped columns from the aggregation will be the indices of the returned object. In order to have the indices restored, we can use reset_index(). The aggregation functions reduce the dimension of the returned object. Moreover, we can apply different functions to different columns:

```
[24]:  df1_groupby.agg({"Result":np.mean,
       "Pulse Duration":np.size})
```

```
[24]:                          Result   Pulse Duration
       Pulse Duration
       10 min                   9.525                4
```

20 min	13.375	4
30 min	17.250	4
5 min	3.925	4

If we want to show a summary of all statistics, one way to do that is by using the `describe` method:

```
[25]: df1_groupby.describe()
```

This leads to the following:

Pulse Duration	Count	Mean	Std	Min	25%	50%	75%	Max
10 min	4.0	9.525	0.377492	9.2	9.200	9.5	9.825	9.9
20 min	4.0	13.375	0.699405	12.5	13.100	13.4	13.675	14.2
30 min	4.0	17.250	0.369685	16.9	16.975	17.2	17.475	17.7
5 min	4.0	3.925	1.426826	2.3	2.975	4.0	4.950	5.4

2. Transformation: Performs some computation to a specific group. This method returns an object which has the same size and index as the grouped object. In the following example we take the grouped object and we select one of the groups to apply two functions to the corresponding values; in this case we compute the square root and an exponential:

```
[26]: df1_groupby.get_group("10 min").transform([np.sqrt,
      np.exp])
```

```
[26]:        Result
          sqrt              exp
      4  3.033150     9897.129059
      5  3.146427    19930.370438
      6  3.130495    18033.744928
      7  3.033150     9897.129059
```

We can also create our own functions and apply them with the transformation method (see example in ◻ Fig. 3.4). Some other built-in useful transformations are (1) `rolling()`, which applies rolling window calculations (there are several window types: Gaussian, Hamming, etc.), and (2) `expanding()`, which accumulates a given operation for all the members of each particular group.

3. Filtration: Discards some groups according to some group criteria. When we apply a function to a group as a filter argument, the output will be Boolean (true or false). The example below groups the `DataFrame df2_melt` by the category `Concentration`. The filter method will then look in the `Result` column for all the values from each concentration group (50 μM, 0.1%...) that have a higher mean value than the overall mean in the column. This is performed by iterating over all the groups mean values of the `DataFrame` using the `lambda`[18] function which will return a true/false value for each of the rows in the filtered column. The printed results will be the ones which were evaluated as true.

[27]:
```
df2_melt.groupby("Concentration").filter(lambda x:
np.mean(x["Result"])>np.mean(df2_melt["Result"]))
```

[27]:

	Treatment	Concentration	Result
3	DMSO	0.5%	100
4	DMSO	0.5%	102
5	DMSO	0.5%	103
9	Drug	50 um	78
10	Drug	50 um	98
11	Drug	50 um	96

`x` is the equivalent to each concentration group from which we compute the mean value from the `Result` column and compare it with the overall mean. As a result, we get that the concentrations of 0.5% of DMSO and 50 μM of Drug have higher mean values than the overall mean.

There are some functions which, when applied to a `DataFrame`, can act as a filter, returning a reduced shape but with unchanged index. For example, when a `Series` or a `DataFrame` are extremely long, but we still want to visualize how the data has been organized in the columns and rows, the functions `head()` and `tail()` come in handy. To view a small sample of a `Series` or a `DataFrame` object, we can use the `DataFrame.head()` method to display the initial (by default, five) values and use the `DataFrame.tail()` method to display the last (by default, five) values.

Combine

The function `pd.concat([df1,df2], axis)` allows to concatenate `DataFrames` along a particular axis; an example is shown in ◘ Fig. 3.4. Once the data is analyzed, we can then combine them into new `DataFrames` and export them into any of the available file formats (using `pd.to_fileformat()`). In the following example, we combine the results from two transformation methods into one new `DataFrame` by using the `pd.concat()` function. Depending on the axis, we can combine the two `DataFrames` horizontally or vertically.

[28]:
```
# We split the data according to the treatment and we
# apply a transformation (a square root  and an
# exponential) to each group: Drug and DMSO
df2_result1 = df2_groupby.get_group("DMSO")
.transform([np.sqrt,f_exp])
df2_result2 = df2_groupby.get_group("Drug")
.transform([np.sqrt,f_exp])
```

18 Lambda functions are commonly used in many programming languages. In Python they allow to create anonymous functions. To learn more about Lambda functions, follow the documentation ▶ https://docs.python.org/3/tutorial/controlflow.

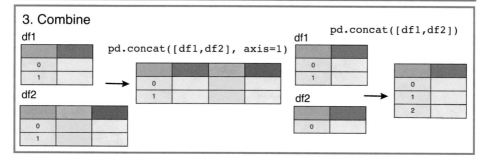

■ **Fig. 3.4** Split-Apply-Combine strategy in Python starting from a tidy data-set `df`. (1) **Split** the data based on some criteria, using the `df.groupby()`. We can then access each of these groups by using the `get_group()` function. (2) **Apply** either an aggregation, a transformation, or a filter operation. Aggregations apply an operation to a group giving one value as a result, such as the mean. Transformations apply a function to all the values of a given group. These functions can be built-in, like `np.exp()` from NumPy, or custom-defined. Filtration applies an operation which returns Boolean indices and, as a result, only the values with true index are shown. Usually the dimensions get reduced from the original size. (3) **Combine** the results using operations like `pd.concat()` to concatenate `DataFrame`, to later export them into CSV or any other table file format

The next step is to combine the results using the concatenation function.

```
[29]: df2_concat = pd.concat([df2_result1, df2_result2],
      axis=0)
      df2_concat
```

```
[29]:          Result
                 sqrt            f_exp
      0       4.472136    4.851652e+08
      1       5.291503    1.446257e+12
      2       5.830952    5.834617e+14
      3      10.000000    2.688117e+43
      4      10.099505    1.986265e+44
      5      10.148892    5.399228e+44
      6       5.656854    7.896296e+13
      7       6.855655    2.581313e+20
      8       7.280110    1.041376e+23
      9       8.831761    7.498417e+33
      10      9.899495    3.637971e+42
      11      9.797959    4.923458e+41
```

We combine now the two results from the Drug treatment and the DMSO treatment, again using the concatenation, but in this case we use the other dimension: pd.concat(axis=1).

```
[30]: df2_concat = pd.concat([df2_result1.rename(index=str,
      columns={"Result": "DMSO"}) ,
      df2_result2.rename(index=str,
      columns={"Result": "Drug"})], axis=1, sort=False)
      df2_concat
```

```
[30]:          DMSO                            Drug
                 sqrt            f_exp         sqrt            f_exp
      0       4.472136    4.851652e+08          NaN             NaN
      1       5.291503    1.446257e+12          NaN             NaN
      2       5.830952    5.834617e+14          NaN             NaN
      3      10.000000    2.688117e+43          NaN             NaN
      4      10.099505    1.986265e+44          NaN             NaN
      5      10.148892    5.399228e+44          NaN             NaN
      6            NaN             NaN     5.656854    7.896296e+13
      7            NaN             NaN     6.855655    2.581313e+20
      8            NaN             NaN     7.280110    1.041376e+23
      9            NaN             NaN     8.831761    7.498417e+33
      10           NaN             NaN     9.899495    3.637971e+42
      11           NaN             NaN     9.797959    4.923458e+41
```

As a result, the two `DataFrames` were horizontally concatenated, but according to their index. Therefore, in order to reset the index, we will use the `df.reset_index(drop=True)` to make the index start from 0 in both `DataFrames`. To avoid the formation of a new column with the index values, we use the `drop=True`.

```
[31]: df2_concat = pd.concat([df2_result1.rename(index=str,
      columns={"Result": "DMSO"}) ,
      df2_result2.reset_index(drop=True).rename(index=str,
      columns={"Result": "Drug"})], axis=1)
      df2_concat
```

```
[31]:          DMSO                          Drug
           sqrt           f_exp          sqrt           f_exp
0     4.472136    4.851652e+08    5.656854    7.896296e+13
1     5.291503    1.446257e+12    6.855655    2.581313e+20
2     5.830952    5.834617e+14    7.280110    1.041376e+23
3    10.000000    2.688117e+43    8.831761    7.498417e+33
4    10.099505    1.986265e+44    9.899495    3.637971e+42
5    10.148892    5.399228e+44    9.797959    4.923458e+41
```

3.4 Python Visualization Landscape

One of the main advantages of using `pandas` data structures, besides the easy handling of the data, is the creation of plots. The structure and metadata inside a `DataFrame` can be easily used to create plots. There is a wide range of different visualization tools available in Python, which should be selected depending on a particular visualization purpose. In this chapter we will focus on Bokeh and HoloViews, JavaScript based packages which produce interactive figures in the browser with the Jupyter Notebook.

3.4.1 JavaScript

JavaScript is a high-level programming language which enables creation of interactive web pages and is frequently used in web applications. Python has many visualization libraries based on JavaScript in order to take advantage of browser interactivity. Currently, having tools which allow easy distribution of the visualization of the data can be very powerful. To learn more about how to turn raw data into interactive web visualizations using a combination of Python and JavaScript, Dale (2016) is a recommended read.

Bokeh

Jupyter Notebook: `NB-3-Bokeh_Plotting.ipynb`

Bokeh is a popular interactive data visualization library for Python which allows to easily share figures. Moreover, Bokeh can handle large and streaming data-sets. To create a figure with Bokeh, the following are the basic steps:

1. Before creating any plot, the first step is to import all the packages and subpackages that will be used to create the figures:

```
[1]:  import pandas as pd
      import numpy as np
      from bokeh.plotting import figure, output_file,
      output_notebook, show
      from bokeh.palettes import Spectral10
```

2. Prepare the data we want to plot, which can be a NumPy array, Python lists, or a pandas DataFrame, as in this example:

```
[2]:  # Define two DataFrames with random numbers
      df1 = pd.DataFrame({"A":np.random.random(60),
      "B":np.random.uniform(0,10,60)})
      df2 = pd.DataFrame({"A":np.random.random(60),
      "B":np.random.uniform(0,10,60)})
      # Choose the columns to be plotted
      x = "A"
      y = "B"
```

3. Define where to generate the output file, using either output_file() (to generate output saved to a file), or output_notebook (to generate output in notebook cells):

```
[2]:  # Specify where to output the figure
      output_notebook()
```

4. Create a figure() object. This will generate a plot with the default options. We can later customize axis labels, title and tools. In this case, we choose some of the most frequently used plot tools which are later explained in more detail in ◘ Fig. 3.5. These tools can be used to zoom-in and -out of the plot, change range extents or to add, edit and delete the graph, etc.

```
[3]:  # Specify the tools if you want to add or remove any
      TOOLS = "crosshair,pan,wheel_zoom,box_zoom,reset,
      box,select,lasso_select"
      # Create a figure
      p = figure(width=400, plot_height=300, tools=TOOLS)
```

5. Add a graph, which can be line(), scatter(), vbar(), hbar(), and many others we can choose from. Some more examples are shown in ◘ Fig. 3.6. Moreover, we can choose color and size of the graphs, label sizes etc.

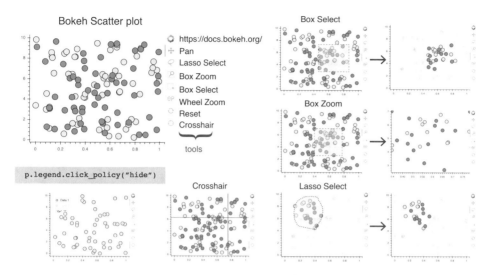

◘ Fig. 3.5 Example of a Bokeh scatter plot and its tools. Two `DataFrames` were used to generate this plot. To the left, there are some of the tools we can activate in the `figure()` section to allow more interactivity. Also, we can have interactive legends which allow you to observe one data-set at a time

```
[4]: # Define the graph
     p.circle(x=x, y=y, source=df1, size=15,
     color=Spectral10[5], line_color="black")
     p.circle(x=x, y=y, source=df2, size=15,
     color=Spectral10[1], line_color="black")
```

6. Choose whether to show the figure, `show(p)` or save it `save(p)`. We cannot generate a vector output like PDFs (Portable Document Formats) or EPS (encapsulated PostScrip) but Bokeh allows us to save in SVG (Scalable Vector Graphics) format.

```
[5]: show(p)
```

The code to generate all of these figures can be found in the Jupyter Notebook: `NB-3-Bokeh_Plotting.ipynb`.

This generates a scatter plot like the one shown in ◘ Fig. 3.5, with the assigned tools. We can add more tools and customize them. Moreover, the plotting parameters can also be adapted, as suitable for a particular figure (visualisation task), but the process always includes all the described basic steps. Some other examples of how to create plots like histograms, box-plots, bar plots and line plots are shown in ◘ Fig. 3.6.

Bokeh has great interactive features. It is a high-level library, but it requires all the described steps to generate a figure. HoloViews will make the process of generating a figure even easier. Their philosophy is: "*Stop plotting your data—annotate your data and let it visualize itself*" (▶ http://holoviews.org).

3

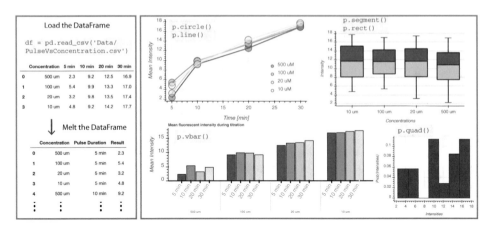

⬛ Fig. 3.6 Examples of different types of Bokeh plots. On the left panel: The `DataFrame` used to generate the plots. This data was already melted and split-up in ▶ Sect. 3.3.5. On the right panel: Some other examples, the code to generate them is in the `NB-3-Bokeh_Plotting.ipynb`

HoloViews

Jupyter Notebook: `NB-4-Holoviews_Plotting.ipynb`

HoloViews is an open-source Python library for simple and easy data analysis and visualization. The approach is that each data-set should have an intrinsic way to be used for its visualization. The intention with this library is to produce intelligent visualizations based on how the data is structured. However, one important point to take into account is that the data must be tidy!

HoloViews can be rendered using either Matplotlib, Bokeh, or Plotly. To do so, we need to specify an extension: `hv.extension("bokeh")` (we will be using Bokeh). Now the plots will be rendered using Bokeh. Next, we generate the figure following the steps below:

1. As before, start by importing the packages needed for creating a figure using HoloViews:

```
[1]:  import pandas as pd
      import numpy as np
      from holoviews import opts
      from bokeh.palettes import Spectral10
      import holoviews as hv
      hv.extension("bokeh")
```

2. Create the data, in this case two `DataFrames`. Specify which columns of the `DataFrames` should be plotted in the figure.

```
[2]:  # Prepare the data you want to plot
      df1 = pd.DataFrame({"A":np.random.random(60),
      "B":np.random.uniform(0,10,60)})
      df2 = pd.DataFrame({"A":np.random.random(60),
```

```
"B":np.random.uniform(0,10,60)})
# Choose the columns you want to plot
x = "A"
y = "B"
```

3. Choose a type of plot (e.g., scatter, box-plot, histogram, heat-map, etc.).

```
[3]: # Type of plot
     scatter1 = hv.Scatter(df1, x, y)
     scatter2 = hv.Scatter(df2, x, y)
```

With these steps we created two objects which will then be rendered with Bokeh (because we chose this extension). Next step is to choose the styling elements for better visualization, using hv.opts().

```
[4]: #Plotting options
     scatter1.opts(color="#fee08b", size=15,
     line_color="black", padding=0.1, tools=TOOLS)
     scatter2.opts(color="#3288bd", size=15,
     line_color="black", padding=0.1, tools=TOOLS)
```

Finally, we choose how we want to visualize the two plots. There are two types of containers: a layout (HoloViews objects displayed side by side, achieved using "+") or an overlay (HoloViews objects displayed overlaid, with the same axes, achieved using "∗").

```
[5]: # Create layout
     scatter1 * scatter2
```

The output from this scatter plot using HoloViews is shown in ◼ Fig. 3.7. As with Bokeh, the plots have, by default, a set of tools which allow more interaction with the data. In this figure there are also some other examples which are explained in more detail in the Jupyter Notebook:
NB-4-Holoviews_Plotting.ipynb.

3

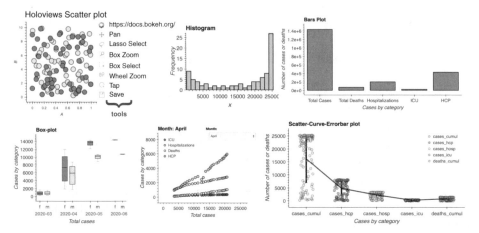

Fig. 3.7 Examples of plots created using HoloViews. The scatter plot corresponds to the step-by-step figure we generated before (◘ Fig. 3.5). The histogram, bar plot, box plot, scatter plot, and scatter-curve-errorbar plot are examples of figures that can be generated using HoloViews (with the Bokeh extension). The plots were generated using a data-set from Covid-19 cases in 2020 in Ireland (▶ https://zenodo.org/record/3901250#.XykEDi17FZI)

Take-Home Message

This chapter provides a guide to use Python as a tool for analyzing and plotting the data as the last step of an image analysis pipeline. With the great increase in number of tools and software to acquire and analyze images, we are able to extract large amount of data (often in the form of tables). pandas is a powerful tool for importing and handling tabular data in Python. However, we need to invest some time to tidy the data in order to get the most out of it when we perform the analysis and the visualization of the results. If we achieve this, we can compute high-level and interactive plots using Bokeh and HoloViews. Tools like the Jupyter Notebooks are very powerful for visualization and data sharing. Utilizing a combination of the interactivity of JavaScript-based visualization libraries (like Bokeh and HoloViews) and efficient handling and analysis tools (like pandas), we can build useful data-analysis pipelines which can be easily shared with others.

Acknowledgements I would like to thank my supervisor, Professor Andrew Charles Oates (EPFL, Lausanne) for the constant support and for the encouragement and freedom to create open science. I would also like to thank Martin Weigert (EPFL, Lausanne) and Uwe Schmidt (CSBD / MPI-CBG) for reviewing this chapter and providing excellent feedback and suggestions. The material for this chapter was initially developed for teaching in courses for NEUBIAS Training schools. I am extremely grateful to the NEUBIAS community for all the discussions and feedback obtained during these meetings. Last but not least, I would like to thank Joan Rué Queralt for reading the chapter and for the great comments.

Further Readings Most of the reading references are provided in the main text. However, for exercises and more examples on this topic, visit ▶ http://bois.caltech.edu. Professor Justin Bois has prepared excellent material for learning data analysis and plotting using Python (specifically, Jupyter notebooks).

References

Dale K (2016) Data visualization with python and javascript: scrape, clean, explore & transform your data. O'Reilly Media, Sebastopol

Harris CR, Millman KJ, van der Walt SJ, Gommers R, Virtanen P, Cournapeau D, Wieser E, Taylor J, Berg S, Smith NJ, Kern R, Picus M, Hoyer S, van Kerkwijk MH, Brett M, Haldane A, del Rìo JF, Wiebe M, Peterson P, G'erard-Marchant P, Sheppard K, Reddy T, Weckesser W, Abbasi H, Gohlke C, Oliphant TE (2020) Array programming with NumPy. Nature 585(7825):357–362. https://doi.org/10.1038/s41586-020-2649-2

McKinney W et al (2011) pandas: a foundational python library for data analysis and statistics. Python High Perform Sci Comput 14(9):1–9

Mertz D (2015) Picking a Python version: a manifesto : from_future_import Python. O'Reilly Media, Sebastopol. https://books.google.ch/books?id=LV74vQEACAAJ

Schindelin J, Arganda-Carreras I, Frise E, Kaynig V, Longair M, Pietzsch T, Preibisch S, Rueden C, Saalfeld S, Schmid B et al (2012) Fiji: an open-source platform for biological-image analysis. Nat Methods 9(7):676–682

Wickham H et al (2011) The split-apply-combine strategy for data analysis. J Stat Softw 40(1):1–29

Wickham H et al (2014) Tidy data. J Stat Softw 59(10):1–23

Building a Bioimage Analysis Workflow Using Deep Learning

Estibaliz Gómez-de-Mariscal, Daniel Franco-Barranco,
Arrate Muñoz-Barrutia and Ignacio Arganda-Carreras

Contents

This Chapter has been reviewed by Sébastien Tosi, IRB, Barcelona.

© The Author(s) 2022
K. Miura, N. Sladoje (eds.), *Bioimage Data Analysis Workflows–Advanced Components and Methods*,
Learning Materials in Biosciences, https://doi.org/10.1007/978-3-030-76394-7_4

What You Will Learn in This Chapter

The aim of this workflow is to quantify the morphology of pancreatic stem cells lying on a 2D polystyrene substrate from phase contrast microscopy images. For this purpose, the images are first processed with a Deep Learning model trained for semantic segmentation (cell/background); next, the result is refined and individual cell instances are segmented before characterizing their morphology. Through this workflow the readers will learn the nomenclature and understand the principles of Deep Learning applied to image processing. Having followed all the steps in this chapter, the reader is expected to know how to use Google Colaboratory (Bisong, 2019) notebooks, ImageJ/Fiji (Rueden et al., 2017; Schindelin et al., 2012; Schneider et al., 2012), DeepImageJ (Gómez-de Mariscal et al., 2019) and MorpholibJ (Legland et al., 2016). This complete workflow sets the basis to develop further methods in the field of Bioimage Analysis using Deep Learning. All the material needed for this chapter is provided in the following GitHub repository (under chap 4): ► https://github.com/NEUBIAS/neubias-springer-book-2021.[1]

4.1 Why You Should Know About Deep Learning

The workflow presented in this Chapter extracts binary masks for cells in 2D phase contrast microscopy images, identifies the cells in the image and quantifies their morphology. The central component of the workflow is the step to obtain a binary mask to distinguish the pixels belonging to the cells from the rest of pixels in the image. In particular, we will train a well established Deep Learning architecture called U-Net (Falk et al., 2019; Ronneberger et al., 2015) to perform this task.

Machine Learning and Deep Learning have become common technical terms in life-science. They are now large fields of study that have boosted both research and industry. While both are strongly related, they also belong to a larger field called Artificial Intelligence, which pursues mimicking (or even surpassing) human intelligence with a machine (Goodfellow et al., 2016). The techniques to extract the proper information and use it in an intelligent way is what we call Machine Learning (ML). The ML techniques are commonly divided into two main groups: supervised and unsupervised methods. Supervised learning is the task of learning a function that maps an input to an output based on sample input-output pairs. Namely, it infers such a function from labeled training data consisting of a set of training examples. When no labels or information about the correct output are given, then we are talking about unsupervised learning, and the corresponding function is inferred using the data structure only. All the clustering methods are thus included in the latter.

A simple example of ML is a linear classifier, technically called *perceptron* (Rosenblatt, 1961), which is able, for example, to split a set of 2D points into two different classes. In practice, ML classifiers operate on objects of way higher dimensions (e.g., images) and solve tasks far more complex than classifying input data into two groups. For this reason, in practice, multiple perceptrons are stacked together to build what is known as an Artificial Neural Network (ANN). That is, we define *deep* architectures to support better mathematical representations of our data. This, combined with a suitable training schedule, allows the computer to learn the correct patterns to per-

---

1 This chapter was communicated by Sébastien Tosi, IRB Barcelona, Spain.

form the desired task. This is called *Deep Learning* (DL from now on) and, at the moment, it has proven to be among the most powerful frameworks for supervised learning.

What sets apart DL from classical approaches is that the system learns automatically from the data without any definition or explicit programming of complex heuristic rules. A pioneer work using DL for bioimage analysis is the Convolutional Neural Network (CNN) architecture called U-Net (Ronneberger et al., 2015). It was first introduced to the community in 2015 at the International Symposium on Biomedical Imaging (ISBI) and then published at the Medical Image Computing and Computer Assisted Interventions (MICCAI) conference, two of the most important conferences for biomedical image analysis. Since then, a growing number of manuscripts (about 390 in 2020 according to PubMed) related to biomedical image analysis using DL are published every year (Litjens et al., 2017).

Note that DL techniques do not only require sophisticated algorithms but also large sets of (manually) annotated images and an enormous amount of computational power. Data collection itself could be a whole project in Computer Vision (Roh et al., 2021), not only for being critical for the success of ML techniques, but also for the complexity that handling large amounts of data involves and the related time and economical costs. In contrast with other fields in Computer Vision, the availability of useful, large and robustly annotated datasets in bioimage analysis is still a bottleneck for the use of DL. This is due to the high economical cost that their acquisition implies, and the need for expertise to generate manual annotations. Indeed, preparing manual annotations can be tedious and many times non-viable. Some freely available annotation tools are QuPath (Bankhead et al., 2017), 3D Slicer (Kapur et al., 2016), Paintera,[2] Mastodon,[3] Catmaid (Saalfeld et al., 2009), TrakEM2 (Cardona et al, 2012), Napari (Sofroniew et al., 2020) and ITK-SNAP (Yushkevich et al., 2006); they offer a wide range of possibilities to simplify the annotation process and make it reasonably efficient. However, there is still a need for a general approach to annotate complex structures in higher dimensions (i.e., 3D, time, multiple channels, multi-modality images). Additionally, the large variability among the images acquired following exactly the same setup but in a different laboratory or by a different technician prevents the transfer of trained DL models. For this reason, we want to warn the reader about the necessity of retraining the DL model provided on the target data to be processed. Fortunately, as it will be demonstrated, this is quite simple to do with a basic knowledge of Python and some libraries such as TensorFlow (Abadi et al., 2016), Keras (Chollet et al., 2015), or Pytorch (Paszke et al., 2019), which release the user from many computational and programming technicalities. Other even more user-friendly frameworks are Ilastik (Berg et al., 2019), ImJoy (Ouyang et al., 2019), ZeroCostDL4Mic (von Chamier et al., 2020), and the ones integrated in Fiji/ImageJ, CSBDeep (Weigert et al., 2018), and deepImageJ (Gómez-de Mariscal et al., 2019). These tools allow the direct use and/or retraining of DL models using zero-code.

(Re)training DL models requires considerable computational power. The use of a graphics processing unit (GPU) such as the ones found in modern graphics boards, or specialized tensor processing units (TPU), is strongly recommended in most cases to speed up the training process. Access to these resources is possible through non-free

---

2 ▸ https://github.com/saalfeldlab/paintera.

3 ▸ https://github.com/mastodon-sc/mastodon.

cloud computing services such as the ones provided by Amazon or Google. Fortunately, there is a free alternative available for Google users through the Google Colaboratory ("Google Colab") framework (Bisong, 2019). It provides serverless Python Jupyter notebooks running on this hardware with pre-installed DL libraries. The use of these resources is limited but most of the time sufficient to train and test bioimage analysis (BIA) models.

4.2 Dataset

The original data processed by this workflow can be found on the web page of the Cell Tracking Challenge (CTC) (Maška et al., 2014; Ulman et al., 2017).[4] It is provided as two independent datasets (training and challenge) since it aims to benchmark (evaluate) cell segmentation and tracking computational methods. The training set is the only one for which Ground Truth[5] (GT) is publicly available. Additionally, the CTC provides a set called Silver Truth[6] (ST). The ST set is much larger than the GT set, so it is more suitable for DL tasks. An example of training data is illustrated in ◘ Fig. 4.1.

For this work, we will use the training set of the challenge and the ST annotations to train and evaluate our method. The ST is processed to extract the contours of each cell that will be used by the workflow (◘ Fig. 4.1). A ready-to-use dataset is provided.[7] Note that the data is distributed into three groups (training, validation and test). We will elaborate more on this in the following sections. For the final step of the workflow, we will apply the trained models to unseen data for which manual

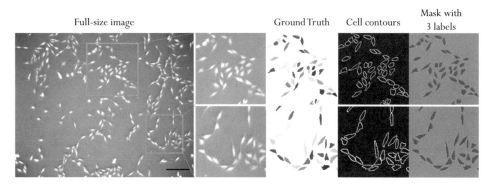

◘ **Fig. 4.1** Example of training data. From left to right: phase contrast microscopy image (scale bar: 150 μm), ground truth (GT) manually annotated cells, corresponding cell-contours, and a mask with 3 labels (background, cell or cell contour)

---

4 ▶ http://celltrackingchallenge.net/2d-datasets/.

5   Ground Truth: It refers to manually annotated images or to the output of controlled simulations. It is the ideal solution that we expect from a computational processing.

6   Silver Truth: It refers to the combination of all the predictions for this particular dataset of the best performing algorithms in the challenge.

7   ▶ https://github.com/NEUBIAS/neubias-springer-book-2021/blob/master/
Ch04_Building_a_Bioimage_Analysis_Workflow_using_Deep_Learning/data4notebooks.zip.

annotations are not available. For this, we will use the challenge data provided at the CTC web page.[8] In a real case scenario, the trained models are always applied to unseen data, with no GT available, otherwise we would not need to train any method!

4.3 Tools

Some tools and software packages need to be installed to run the workflow:
- Fiji[9]
  - Download URL: ► https://imagej.net/Fiji/Downloads
  - MorphoLibJ plugin. IJPB update site URL: ► https://sites.imagej.net/IJPB-plugins/
  - DeepImageJ plugin. Update site URL: ► https://sites.imagej.net/DeepImageJ/
  To install Fiji plugins, in Fiji, click on `Help > Update...` Once the ImageJ Updater opens, click on `Manage update sites`. There you need to select the IJPB-plugins for MorpholibJ. To install deepImageJ, you need to click on `Add update site`. Then, fill the fields with Name: DeepImageJ and update site URL. Click on `Close` and `Apply changes`.
- Python Notebooks: they can be executed locally or in Google Colaboratory[10] which provides free access to cloud GPU. The latter requires a Google account.
  - Link to the notebook.[11]
  - Link to open the notebook directly in Google Colaboratory.[12] It is recommended to make a local copy of the Notebook, as it will be editable.

4.4 Workflow

The steps of the workflow covered in this chapter are summarized in ◘ Fig. 4.2.

4.4.1 Step 1: Setting up a Google Colaboratory Notebook

After opening a Google Colab notebook, we configure the hardware needed for its execution. In this case, we set up a GPU runtime (◘ Fig. 4.3). Now we can run the notebook. The way to proceed is by clicking on the "play" button on the left side of each code cell. For example, the first cell will install the correct version of the required

---

8 ► http://data.celltrackingchallenge.net/challenge-datasets/PhC-C2DL-PSC.zip.

9 All the steps described in this chapter are reproducible in Fiji and ImageJ.

10 ► https://colab.research.google.com/.

11 ► https://github.com/NEUBIAS/neubias-springer-book-2021/blob/master/
   Ch04_Building_a_Bioimage_Analysis_Workflow_using_Deep_Learning/notebook/
   U_Net_PhC_C2DL_PSC_segmentation.ipynb.

12 ► https://colab.research.google.com/github/NEUBIAS/neubias-springer-book-2021/blob/
   master/Ch04_Building_a_Bioimage_Analysis_Workflow_using_Deep_Learning/notebook/
   U_Net_PhC_C2DL_PSC_segmentation.ipynb.

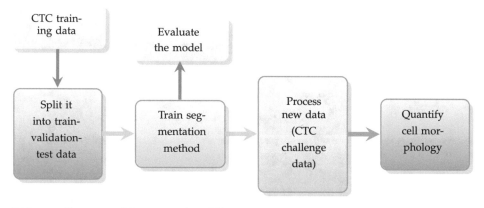

◘ Fig. 4.2 Summary of the proposed workflow

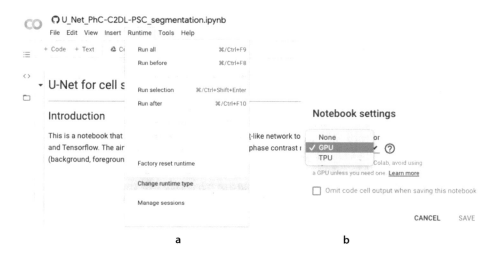

a   b

◘ Fig. 4.3 Setting up a Google Colab notebook. (a) Go to "Change runtime type" and (b) make sure to choose GPU hardware

DL libraries (TensorFlow and Keras). This is critical for results reproducibility since functions performance can differ among different versions, or the code may even crash (◘ Fig. 4.4).

4.4.2 Step 2: Download and Split the Data into Training, Validation and Test

When using ML methods, we need to split the available annotated (GT) data into three exclusive sets: training, validation and test. The training set is used to train the method and let it learn the task of interest (e.g., binary segmentation). Such set needs to be large enough as to cover all representative scenarios (e.g., poor signal-to-noise ratio, blurred images) and events visible in the data (e.g., artifacts, debris, mitosis,

▾ Getting started

First, we make sure we are using Tensorflow version compatible with DeepImageJ (<= 1.15).

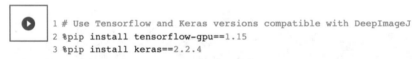

```
1 # Use Tensorflow and Keras versions compatible with DeepImageJ
2 %pip install tensorflow-gpu==1.15
3 %pip install keras==2.2.4
```

■ **Fig. 4.4**   Execution of the first code cell. Every piece of code is run by clicking on the play button (red square) of each code cell

apoptosis, clusters of cells). The validation set, as indicated by its name, serves to evaluate the performance of the method during training, to ensure that it is learning and to prevent over-fitting.[13] The test set will be used to assess the performance of the method once the training procedure has finished. Both validation and test sets need to be independent of the training set, so that when the accuracy of the model becomes acceptable on the validation set, we can be confident that it is because the model is properly trained and that it has not over-fit the training set. The evaluation of the model performance on the test set aims to assess its ability to generalize to unseen data.

The GT data, in this particular case, consists of two independent time-lapse videos (sequences 01 and 02). Some frames from sequence 01 are used as training data while some other frames from the sequence 02 are used for both validation (frames 140, …, 250) and test (frames 151, 152, …, 248, 249). This data organization is compiled in a zip file that needs to be downloaded and unzipped (in the cloud, if running the workflow in Google Colab). These operations are performed in the second code cell by the following commands:

```
1  import zipfile
2  # Download file with image data
3  !(wget 'https://github.com/NEUBIAS/neubias-springer-book-2021/raw/master/\
4  Ch04_Building_a_Bioimage_Analysis_Workflow_using_Deep_Learning/\
5  data4notebooks.zip';)
6  path2zip= 'data4notebooks.zip'
7  # Extract locally
8  with zipfile.ZipFile(path2zip, 'r') as zip_ref:
9      zip_ref.extractall('/content/dataset/')
```

code cell[1], U_Net_PhC_C2DL_PSC_segmentation.ipynb

After decompression, the new folder called `dataset` contains three sub-folders (`input`, `binary_masks` and `contours`) for the three different sets.

4.4.3   Step 3: Train a Deep Learning Model for Binary Segmentation

A U-Net DL network is designed and trained to segment the cells in the images. We train the network by using the original 2D phase contrast microscopy images as input, and a set of three binary masks as output: 1) background mask (with pixel values of 1 for the background and 0 for the rest), 2) cell mask (1 cells and 0 the rest) and 3)

---

13   When the model processes the training data accurately but fails to generalize the accurate prediction to the test set, we say that it over-fits the training data.

cell contour (1 cell contour and 0 the rest). In other words, the network will learn to classify each input pixel as belonging to one of these three classes: *background, foreground* or *contour*.

Since the classification is performed per pixel, this process is called *semantic segmentation*, as opposed to *instance segmentation*, for which the model outputs a unique label per object of interest (here, independent cells).

Step 3.1: Preparing the Data for Training

Read the images for training and store them into memory by running the following code:

```
1   # Path to the training images
2   train_input_path = '/content/dataset/train_input'
3   train_masks_path = '/content/dataset/train_binary_masks'
4   train_contours_path = '/content/dataset/train_contours'
5   # Read the list of file names and sort them to have a match between images and masks
6   train_input_filenames = [x for x in os.listdir( train_input_path ) if x.endswith(".tif")]
7   train_input_filenames.sort()
8   train_masks_filenames = [x for x in os.listdir( train_masks_path ) if x.endswith(".tif")]
9   train_masks_filenames.sort()
10  train_contours_filenames = [x for x in os.listdir( train_contours_path ) if x.endswith(".png")]
11  train_contours_filenames.sort()
12  print( 'Number of training input images: ' + str( len(train_input_filenames)) )
13  print( 'Number of training binary mask images: ' + str( len(train_masks_filenames)) )
14  print( 'Number of training contour images: ' + str( len(train_contours_filenames)) )
15  # Read training images (input, mask and contours)
16  train_img = [cv2.imread(os.path.join(train_input_path, x), cv2.IMREAD_ANYDEPTH) for x in train_input_filenames ]
17  train_masks = [cv2.imread(os.path.join(train_masks_path, x), cv2.IMREAD_ANYDEPTH)>0 for x in train_masks_filenames ]
18  train_contours = [cv2.imread(os.path.join(train_contours_path, x), cv2.IMREAD_ANYDEPTH)>0 for x in
↪    train_contours_filenames ]
19  # display the image
20  plt.figure(figsize=(10,5))
21  plt.subplot(1, 3, 1)
22  plt.imshow( train_img[0], 'gray' )
23  plt.axis('off')
24  plt.title( 'Full-size training image' )
25  # its "mask"
26  plt.subplot(1, 3, 2)
27  plt.imshow( train_masks[0], 'gray' )
28  plt.axis('off')
29  plt.title( 'Binary mask' )
30  # and cell contours
31  plt.subplot(1, 3, 3)
32  plt.imshow( train_contours[0], 'gray' )
33  plt.axis('off')
34  plt.title( 'Object contour' )
35  # Concatenate binary masks and contours to get one array with the training data
36  train_output = [np.transpose(np.array([train_masks[i],train_contours[i]]), [1,2,0]) for i in range(len(train_masks))]
```

code cell[2-4], U_Net_PhC_C2DL_PSC_segmentation.ipynb

You should get the following message together with the figures from ▢ Fig. 4.5.

```
1   Number of training input images: 101
2   Number of training binary mask images: 101
3   Number of training contour images: 101
```

output of code cell[2-4], U_Net_PhC_C2DL_PSC_segmentation.ipynb

The U-Net network we are going to train has $\sim 500,000$ trainable parameters, which requires a large amount of memory. Thus, to reduce memory usage and make it fit to the hardware offered by Google Colab, we crop small random patches of size 256×256 pixels from the original images. To do so, we create a function that crops a fixed number of patches from each image. We need to make sure that the part cropped out from the input image and the output patches (annotation binary masks) correspond to each other. Then, we use this function to crop out patches from the training data in the following code section:

Full-size traning image   Binary mask   Object contour

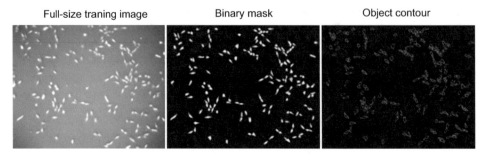

▣ Fig. 4.5 Output of "Preparing the data for training" code section displaying one training image and corresponding annotations

```
1   def create_random_patches( imgs, masks, num_patches, shape ):
2       ''' Create a list of image patches out of a list of images
3       Args:
4           imgs (list): input images.
5           masks (list): binary masks (output images) corresponding to imgs.
6           num_patches (int): number of patches for each image.
7           shape (2D array): size of the patches. Example: [256, 256].
8       Returns:
9           list of image patches and patches of corresponding labels (background,
10          foreground and contours)
11      '''
12      original_size = imgs[0].shape
13      input_patches = []
14      output_patches = []
15      for n in range( 0, len( imgs ) ):
16          image = imgs[ n ]
17          mask = masks[ n ]
18          for i in range( num_patches ):
19              r = np.random.randint(0,original_size[0]-shape[0])
20              c = np.random.randint(0,original_size[1]-shape[1])
21              input_patches.append( image[ r : r + shape[0], c : c + shape[1] ] )
22              output_patches.append( mask[ r : r + shape[0], c : c + shape[1] ] )
23      return input_patches, output_patches
24  # Use the method to create six 256x256 pixel-sized patches per image
25  train_input_patches, train_output_patches = create_random_patches(train_img,train_output,6, [256,256])
26  # In X_train we will store the input images
27  X_train = [x/255 for x in train_input_patches] # normalize between 0 and 1
28  X_train = np.expand_dims(X_train, axis=-1)
29  print('There are {} patches to train the network'.format(len(X_train)))
```

```
part of code cell[7], U_Net_PhC_C2DL_PSC_segmentation.ipynb
```

We choose to normalize the intensity values of the input and output images between 0.0 and 1.0. This way, a common range of values for all the images is set without changing the differences among them or their properties. This helps the network to find the optimal parameters which give generality to the model and in some cases, to speed up the training.

Note that the class of each pixel is mathematically written using a *one-hot encoding* representation, for which we need three binary matrices (one per class) for each image. Hence, a pixel in the background is encoded as [1, 0, 0], as [0, 1, 0] for foreground and as [0, 0, 1] for cell contour. This is performed by the following code section:

```
1   # In Y_train we will store the target labels for the network in a one-hot fashion, so first channel for background,
    ↪   second for foreground (cells) and third for cell boundaries (cell contours)
2   Y_train = [np.stack([1 - x[:,:,0] - x[:,:,1], x[:,:,1], x[:,:,0]],
3               axis=-1) for x in train_output_patches ]
4   Y_train = np.asarray( Y_train )
```

```
part of code cell[7], U_Net_PhC_C2DL_PSC_segmentation.ipynb
```

Exercise 1

Repeat the same procedure for the validation set. You should obtain two variables X_val and Y_val with shapes $n \times 256 \times 256 \times 1$ and $n \times 256 \times 256 \times 3$, respectively, n being the total number of patches generated from the validation set. We recommend to generate 6 patches for each image as there are only 11 images in the validation set and you will only crop small patches from them.

Step 3.2: Building a U-Net Shaped Convolutional Neural Network

The key component of any DL method used for image analysis are the convolutional layers: A filter kernel, convolution matrix, which is a small matrix that is convolved with the input image (see ◘ Fig. 4.6a). Convolution is a (linear) operation of summing elements in a local neighbourhood in the image, each weighted by the given kernel coefficients, with an aim to cause an effect on the input image (i.e., blurring, enhancement, edge detection). In the DL context, we use the word kernel when referring to this small matrix. The coefficients of the matrix are called the kernel weights. The learning process consists of finding the optimal weights for each convolutional kernel. Most of the time, the features extracted with the convolutional layers are not complex enough as to represent and analyze the relevant information in the image. A common strategy is to encode the features into a high dimensional space, process them and recover the original spatial representation by decoding the processed features. In the encoding path, the number of filters in the convolutional layer is increased and the size of the image decreased. This way, a higher dimensional space of features is reached (see ◘ Fig. 4.6b). To recover the original spatial representation, the number of filters is decreased as the spatial dimensions are increased (see ◘ Fig. 4.6d). The architectures that follow this schema are called encoder-decoders. A well established encoder-decoder for biomedical image analysis is the U-Net, which has encoding levels in the contracting path (the encoder), a bottleneck and decoding levels in the expanding path (decoder). See ◘ Fig. 4.7 for a graphical description of the U-Net-like architecture used in the current workflow.

The layers in Keras can be defined as `output = Operation(number of filters, size)(input)`. Some additional arguments that can be specified are: the type of activation function used in the convolutional layer (`activation`), the initial distribution of the weights (`kernel_initializer`), and whether to use zero padding or not to preserve the size of the images after every convolution (`padding`).

The encoding path of the U-Net can be programmed simply by a downsampling of the image. Here we use `AveragePooling2D`.[14] Similarly, the decoding can be achieved by upsampling. However, in this case, we decided to use transposed or inversed convolutions (`Conv2DTranspose`) that need to be trained as well as the convolutional layers. The final configuration is as follows:

```
1  # We leave the height and width of the input image as "None" so the network can
2  # later be used on images of any size.
3  inputs = Input((None, None, 1))
4  # Contracting path
5  c1 = Conv2D(16, (3, 3), activation='elu', kernel_initializer='he_normal', padding='same') (inputs)
6  c1 = Dropout(0.1) (c1)
7  c1 = Conv2D(16, (3, 3), activation='elu', kernel_initializer='he_normal', padding='same') (c1)
8  p1 = AveragePooling2D((2, 2)) (c1)
9
```

14   More pooling layer types at ▶ https://keras.io/api/layers/pooling_layers/.

4

a

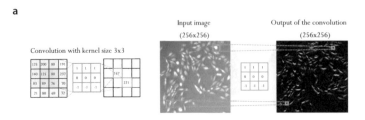

b

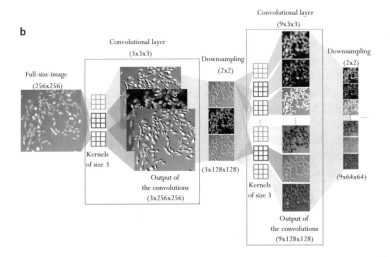

c

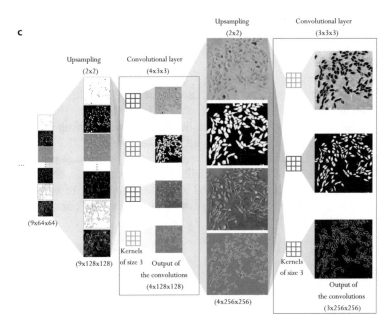

☐ **Fig. 4.6** (**a**) Convolution of an image using a kernel of size 3 × 3. (**b**) 2 level encoding of an input image into a feature space using convolutions and downsamplings. (**c**) 2 level decoding of a set of features into the original spatial dimension. In (**b**) and (**c**), the convolutional layers have 3 and 9, and 4 and 3 filters, respectively. All the kernels have size 3 × 3 and their weights are trainable parameters that are optimized during the training. Downsampling and upsampling have size 2 × 2, so the image size is halved and doubled, respectively

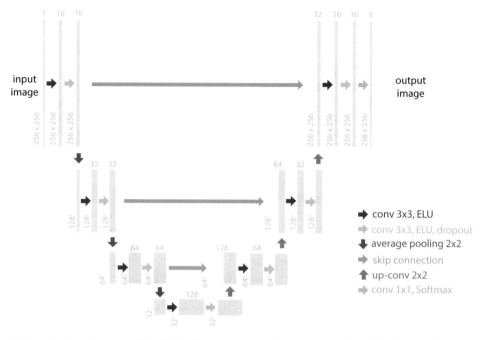

Fig. 4.7 Architecture of the U-Net-like convolutional neural network used in the workflow

```
10  c2 = Conv2D(32, (3, 3), activation='elu', kernel_initializer='he_normal', padding='same') (p1)
11  c2 = Dropout(0.2) (c2)
12  c2 = Conv2D(32, (3, 3), activation='elu', kernel_initializer='he_normal', padding='same') (c2)
13  p2 = AveragePooling2D((2, 2)) (c2)
14
15  c3 = Conv2D(64, (3, 3), activation='elu', kernel_initializer='he_normal', padding='same') (p2)
16  c3 = Dropout(0.3) (c3)
17  c3 = Conv2D(64, (3, 3), activation='elu', kernel_initializer='he_normal', padding='same') (c3)
18  p3 = AveragePooling2D((2, 2)) (c3)
19
20  # Bottleneck
21  c4 = Conv2D(128, (3, 3), activation='elu', kernel_initializer='he_normal', padding='same') (p3)
22  c4 = Dropout(0.4) (c4)
23  c4 = Conv2D(128, (3, 3), activation='elu', kernel_initializer='he_normal', padding='same') (c4)
24
25  # Expanding path
26  u5 = Conv2DTranspose(64, (2, 2), strides=(2, 2), padding='same') (c4)
27  u5 = concatenate([u5, c3])
28  c5 = Conv2D(64, (3, 3), activation='elu', kernel_initializer='he_normal', padding='same') (u5)
29  c5 = Dropout(0.3) (c5)
30  c5 = Conv2D(64, (3, 3), activation='elu', kernel_initializer='he_normal', padding='same') (c5)
31
32  u6 = Conv2DTranspose(32, (2, 2), strides=(2, 2), padding='same') (c5)
33  u6 = concatenate([u6, c2])
34  c6 = Conv2D(32, (3, 3), activation='elu', kernel_initializer='he_normal', padding='same') (u6)
35  c6 = Dropout(0.2) (c6)
36  c6 = Conv2D(32, (3, 3), activation='elu', kernel_initializer='he_normal', padding='same') (c6)
37
38  u7 = Conv2DTranspose(16, (2, 2), strides=(2, 2), padding='same') (c6)
39  u7 = concatenate([u7, c1], axis=3)
40  c7 = Conv2D(16, (3, 3), activation='elu', kernel_initializer='he_normal', padding='same') (u7)
41  c7 = Dropout(0.1) (c7)
42  c7 = Conv2D(16, (3, 3), activation='elu', kernel_initializer='he_normal', padding='same') (c7)
43
44  # The output will consist of 3 neurons (one per class) with softmax activation
45  # so they represent probabilities
46  outputs = Conv2D(3, (1, 1), activation='softmax') (c7)
47
48  model = Model(inputs=[inputs], outputs=[outputs])
49  model.summary()
```

part of code cell[10], U_Net_PhC_C2DL_PSC_segmentation.ipynb

Note that the layers are sequentially connected, that is, the output of a layer is the input of the following layer.

Step 3.3: Loss and Accuracy Measures

The training schedule is a common optimization process. During each iteration of the training, the output of the CNN is compared with the corresponding GT through a loss function (summarizing the differences between them as a numerical value). Hence, the learning process consists in minimizing the loss function. To perform this optimization, the gradient of the loss function is computed and the network parameters (the kernels weights) are updated accordingly in the direction of the gradient variation by step sizes proportional to the learning rate.

The most common loss functions are the mean squared error (MSE), the binary cross-entropy (BCE) and the categorical cross-entropy (CCE). MSE is used for regression problems (when the output is not a class but a continuous value), while BCE and CCE are used in classification tasks. Patterson and Gibson (2017) provide further details about loss functions in DL. TensorFlow and Keras have also implemented quite many ready-to-use loss functions.[15] Standard optimizers for neural networks are the Stochastic Gradient Descent (SGD) (Kiefer et al., 1952), Root Mean Square propagation (RMSprop)[16] and Adaptive Moment Estimation (Adam) (Kingma and Ba, 2014). The latter is an optimization algorithm specifically designed for DL.

Here, we use the CCE loss function (Eq. 4.1), and the Adam optimizer with a learning rate set to 0.0003 (experimentally estimated but learning rates are typically in this range of values; see comments in Appendix):

$$CCE(y, p) = - \sum_{c=1}^{C} y_{i,c} log(p_{i,c}) \tag{4.1}$$

where y is the GT, p the predicted value, C the total number of classes ($C = 3$ in this case); $y_{i,c} = 1$ if the class of the observation i is c and 0, otherwise, and $p_{i,c}$ is the predicted probability for the observation i of being of class c. The values of the loss function are usually difficult to interpret since the better the performance is, the lower its value. The accuracy measure gives an indication of how close is the output of the network to the Ground Truth. This metric is easier to interpret and visualize than the loss value but it is not suitable to guide the network optimization during training. Its values are limited to the [0, 1] range, 1 being a perfect match between the result and the GT. Some standard accuracy measures for classification are the Jaccard index (also called Intersection over Union (IoU)), the Dice coefficient, the Hausdorff distance and the rate of True or False Positives and Negatives.

In Keras, many standard loss functions are available but we need to define a suitable accuracy measure for the problem at hand. As we deal with a segmentation

---

15 ▶ https://www.tensorflow.org/api_docs/python/tf/keras/losses.

16 G. Hinton, 2012 (▶ https://www.cs.toronto.edu/~tijmen/csc321/slides/lecture_slides_lec6.pdf).

task, we will use the Jaccard index, a good indicator of the overlap between our predicted and target segmented cells. It is defined for a binary image as:

$$J(y, p) = \frac{|y \cap p|}{|y \cup p|} = \frac{TP}{TP + FN + FP} \tag{4.2}$$

where y is the GT, p the predicted value, TP the true positives, FN false negatives and FP false positives. Note that the Jaccard index measures the ratio of correctly classified pixels. Although the network output has three channels (background, foreground and object-contours), we compute the accuracy measure as the average Jaccard index of the last two classes (channels). Since many pixels belong to the background class, including them into the computation would produce misleadingly high Jaccard index values. A function computing this metric can be implemented in TensorFlow as follows:

```
1   def jaccard_index(y_true, y_pred, skip_background=True):
2       """Define Jaccard index for multiple labels.
3       Args:
4           y_true (tensor): ground truth masks.
5           y_pred (tensor): predicted masks.
6           skip_background (bool, optional): skip 0-label from calculation
7       Return:
8           jac (tensor): Jaccard index value
9       """
10      # We read the number of classes from the last dimension of the true labels
11      num_classes = tf.shape(y_true)[-1]
12      # one_hot representation of predicted segmentation after argmax
13      y_pred_ = tf.one_hot(tf.math.argmax(y_pred, axis=-1), num_classes)
14      y_pred_ = tf.cast(y_pred_, dtype=tf.int32)
15      # y_true is already one-hot encoded
16      y_true_ = tf.cast(y_true, dtype=tf.int32)
17      # skip background pixels from the Jaccard index calculation
18      if skip_background:
19          y_true_ = y_true_[...,1:]
20          y_pred_ = y_pred_[...,1:]
21      TP = tf.math.count_nonzero(y_pred_ * y_true_)
22      FP = tf.math.count_nonzero(y_pred_ * (y_true_ - 1))
23      FN = tf.math.count_nonzero((y_pred_ - 1) * y_true_)
24      jac = tf.cond(tf.greater((TP + FP + FN), 0), lambda: TP / (TP + FP + FN),
25                    lambda: tf.cast(0.000, dtype='float64'))
26      return jac
```

code cell[9], U_Net_PhC_C2DL_PSC_segmentation.ipynb

Once the network and all the required functions have been defined, we can compile the model by calling:

```
1   # Finally compile the model with Adam as optimizer, CCE as loss function and Jaccard as metric
2   opt = keras.optimizers.Adam(lr=0.0003) # Adam with specified learning rate
3   model.compile(optimizer=opt, loss='categorical_crossentropy', metrics=[jaccard_index])
```

part of code cell[10], U_Net_PhC_C2DL_PSC_segmentation.ipynb

Step 3.4: Executing the Training Schedule

We set up the training schedule with a maximum of 100 epochs[17] and a batch size[18] of 10. The validation accuracy is monitored during the training. If it does not change for a certain number of epochs (i.e., patience), then the training process is interrupted and the best performing instance of the model is returned. Patience is initially set to 50 using the `EarlyStopping` callback of Keras.

---

17 Epochs: the number of times that the whole data is covered in the learning process.
18 Batch size: the number of training examples seen by the network before updating its weights.

To execute the training process, we just need to specify the training (X_train and Y_train) and the validation data (X_val and Y_val). During the training, the model (variable model) is automatically updated:

```
1   # Training parameters
2   numEpochs = 100 # maximum number of epochs to train
3   patience = 50   # number of epochs to wait before stopping if no improvement
4   batchSize = 10  # number of samples per batch
5   # Define early stopper to finish the training when the network does not improve
6   earlystopper = EarlyStopping(patience=patience, verbose=1, restore_best_weights=True,
7                                monitor='val_jaccard_index', mode='max')
8   # Train!
9   history = model.fit( X_train, Y_train, validation_data = (X_val, Y_val),
10                       batch_size = batchSize, epochs=numEpochs,
11                       callbacks=[earlystopper])
12  # # Save the model weights to an HDF5 file
13  model.save_weights( 'unet_pancreatic_cell_segmentation_best.h5' )
14  >>>> Output of the code
15  606/606 [==============================] - 27s 45ms/step - loss: 0.4661 - jaccard_index: 0.0060 - val_loss: 0.2805 -
    ↪ val_jaccard_index: 0.0027
16  Epoch 2/100
17  606/606 [==============================] - 15s 24ms/step - loss: 0.2581 - jaccard_index: 0.3045 - val_loss: 0.1572 -
    ↪ val_jaccard_index: 0.4238
18  ...
19  Epoch 100/100
20  606/606 [==============================] - 15s 24ms/step - loss: 0.0395 - jaccard_index: 0.8098 - val_loss: 0.0539 -
    ↪ val_jaccard_index: 0.7784
```

code cell[11], U_Net_PhC_C2DL_PSC_segmentation.ipynb

It is possible to store the details of the training for each epoch (variable history in the code) and plot them afterwards (▣ Fig. 4.8):

```
1   plt.figure(figsize=(14,5))
2   # summarize history for loss
3   plt.subplot(1, 2, 1)
4   plt.plot(history.history['loss'])
5   plt.plot(history.history['val_loss'])
6   plt.title('model loss')
7   plt.ylabel('loss')
8   plt.xlabel('epoch')
9   plt.legend(['train', 'val'], loc='upper left')
10  # summarize history for Jaccard index
11  plt.subplot(1, 2, 2)
12  plt.plot(history.history['jaccard_index'])
13  plt.plot(history.history['val_jaccard_index'])
14  plt.title('model Jaccard index')
15  plt.ylabel('Jaccard index')
16  plt.xlabel('epoch')
17  plt.legend(['train_jacc', 'val_jacc'], loc='lower right')
18  plt.show()
```

code cell[12], U_Net_PhC_C2DL_PSC_segmentation.ipynb

In ▣ Fig. 4.8, we can observe that the loss value in the training dataset decreases after each epoch while the loss for the validation data does only decrease until epoch 40 and then starts to increase slightly. This is a sign that the training cannot further improve the model and could even degrade it by over-fitting to the training dataset. A similar behavior can be observed when looking at the Jaccard index. It seems that the method can still do it better for the training dataset but not for the validation set. This is the second hint pointing that the model was optimized as much as possible given the training data.

❓ Exercise 2

Train the network using a smaller amount of images. This can be done easily, by reducing the file lists train_input_filenames, train_masks_filenames and train_contours_filenames, in Step 3.1. You will notice that when using few images the accuracy of the network on the validation and test data is decreased. We suggest to increase the number of epochs so you can also visualize any existing over-fitting or whether the network needs a longer training process.

4.4.4 **Step 4: Evaluating the Trained Model**

Keras enables simple evaluation of the performance of the method as long as the same information as for the training is available for the test dataset (input and GT images). For this, we just need to initialize two variables X_test and Y_test, see Exercise 3.

```
1   # Evaluate the model on the test data using `evaluate`
2   results = model.evaluate(X_test, Y_test , batch_size=1)
3   print('test loss CCE: {0}, Jaccard index: {1}'.format(results[0], results[1]))
4   >>>> Output of the code
5   90/90 [==============================] - 11s 118ms/step
6   test loss CCE: 0.09035193290975359, Jaccard index: 0.7407998955328148
```

code cell[16], U_Net_PhC_C2DL_PSC_segmentation.ipynb

> **❓ Exercise 3**
>
> Same as what was asked in Exercise 1, read the images in the test folder and create two normalized Numpy arrays X_test and Y_test. However, note that random patches are not adopted this time as we want to evaluate the performance on the whole image. Additionally, the size of the network input needs to be a multiple of 16 due to the downsampling layers and skip connections (■ Fig. 4.7). Hence, crop the largest possible (560×704 pixels) central patch for each image and its manual annotations. The expected shapes of X_test and Y_test are $90 \times 560 \times 704 \times 1$ and $90 \times 560 \times 704 \times 3$, respectively.

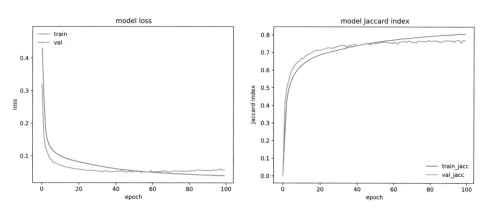

■ **Fig. 4.8**   Plotting the training loss and Jaccard index per epoch. The training was set to 100 epochs and values stored in the variable history are displayed. Two metrics are calculated: Categorical Cross Entropy (CCE) and Jaccard index, as loss and accuracy. The values for the training data are shown in blue, and for validation in orange

4.4.5 Step 5: Building a DeepImageJ Bundled Model to Process New Data

Step 5.1: Saving the Trained Model in TensorFlow's Format

DeepImageJ is a plugin toolset in Fiji/ImageJ designed to load and run TensorFlow models. Next, we show how to store the model in a SavedModel ProtoBuffer format (default file format in TensorFlow), so that deepImageJ can read it and process an image directly loaded from ImageJ using the trained model:

```
1   # Folder in which the model is stored. This folder must not exist.
2   OUTPUT_DIR = "DeepImageJ-model"
3   builder = tf.saved_model.builder.SavedModelBuilder(OUTPUT_DIR)
4   signature = tf.saved_model.signature_def_utils.predict_signature_def(
5           inputs  = {'input':  model.input},
6           outputs = {'output': model.output})
7   signature_def_map = { tf.saved_model.signature_constants.DEFAULT_SERVING_SIGNATURE_DEF_KEY: signature }
8   builder.add_meta_graph_and_variables(K.get_session(), [tf.saved_model.tag_constants.SERVING],
    ↪   signature_def_map=signature_def_map)
9   builder.save()
```

code cell[18], U_Net_PhC_C2DL_PSC_segmentation.ipynb

A new folder called *DeepImageJ-model* is created with two items inside: *saved_model.pb* and a folder *variables*. We recommend to compress this folder into a `DeepImageJ-model.zip` file and download it so you can work on it locally with Fiji/ImageJ:

```
1   from google.colab import files
2   !zip DeepImageJ-model -r DeepImageJ-model/
3   # Download!
4   files.download("DeepImageJ-model.zip")
```

code cell[23], U_Net_PhC_C2DL_PSC_segmentation.ipynb

Unzip the file in your local machine. Note that the folder should look exactly like the one we had in the cloud (*DeepImage-model*).

Step 5.2: Creating a DeepImageJ Bundled Model

DeepImageJ comprises three different plugins: Run, Explore and BuildBundled Model. First, the TensorFlow model needs to be converted into a deepImageJ's bundled model. Click on *ImageJ > Plugins > DeepImageJ > Build BundledModel* and open an example image for this processing. We opened the image t199.tif from the test set. A dialog box pops up indicating the steps to follow (see ◘ Fig. 4.9).

The pre-processing ImageJ macro[19] is used to normalize the input images:

```
1   // Preprocessing macro
2   print("Preprocessing");
3   run("32-bit");
4   getRawStatistics(nPixels, mean, min, max, std, histogram);
5   run("Divide...", "value=" + max);
```

ImageJ macro for pre-processing in DeepImageJ

If no post-processing macro is set, we get the raw output of the network (◘ Fig. 4.10). However, we would like to identify each independent cell in the mask (i.e., *instance*

---

19 Macro available at ▸ https://github.com/NEUBIAS/neubias-springer-book-2021/tree/master/Ch04_Building_a_Bioimage_Analysis_Workflow_using_Deep_Learning/ij-macros/preprocessing.txt.

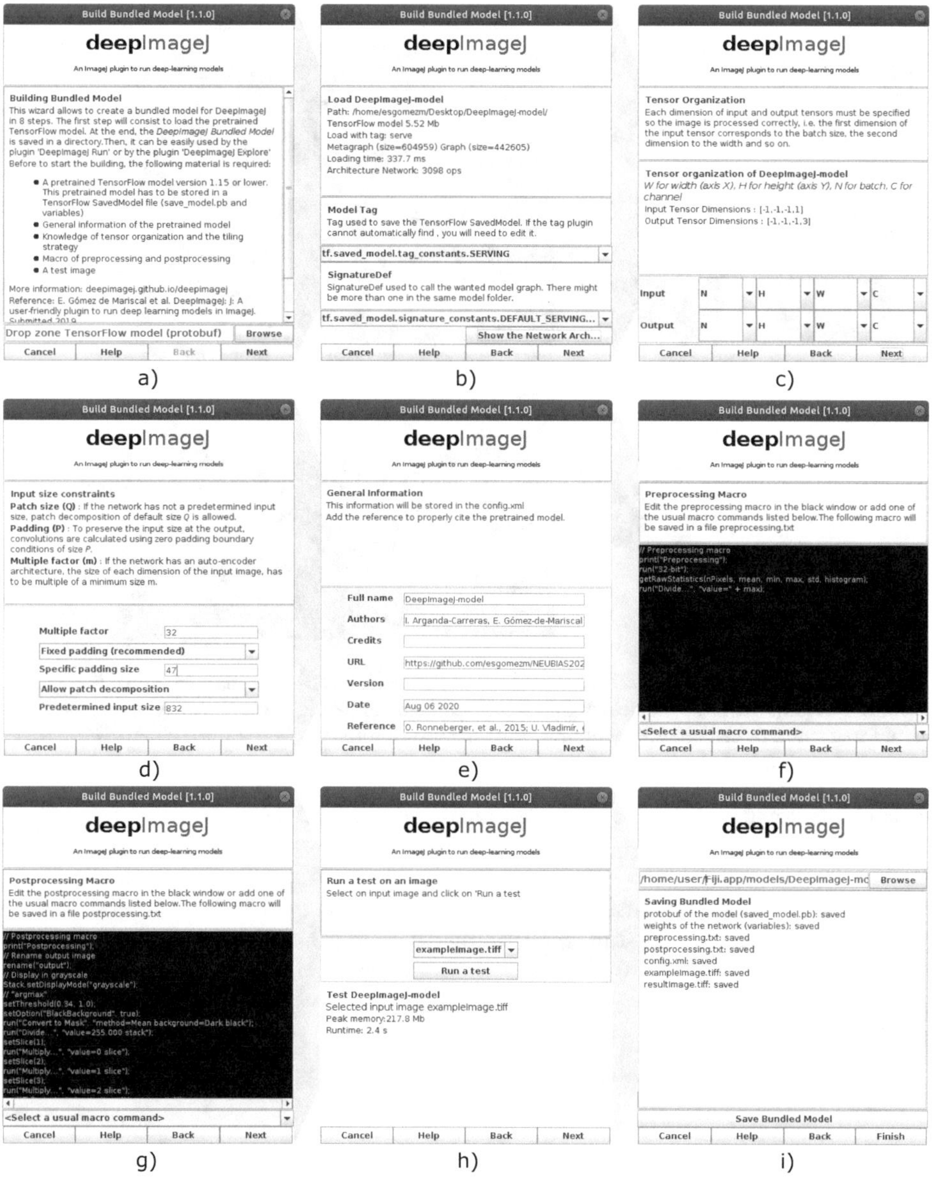

Fig. 4.9 DeepImageJ build bundled model process: (**a**) Open a test image in Fiji and call *Build Bundled Model*; (**b**) Load a model indicating the path to the unzipped *DeepImageJ-model* folder; (**c**) Specify input and output dimension order (N: batch number, H: height, W: width, C: channels); and also (**d**) input size (32) and padding (47); (**e**) Write the name of the model, authors, credits, citations or any other relevant information; (**f, g**) Write the pre- and post-processing macro routines needed for the correct image processing; (**h**) Run the image processing routine and test that you get the desired output; (**i**) If so, specify a new name for the bundled model and save it under ImageJ's recently created *models* folder

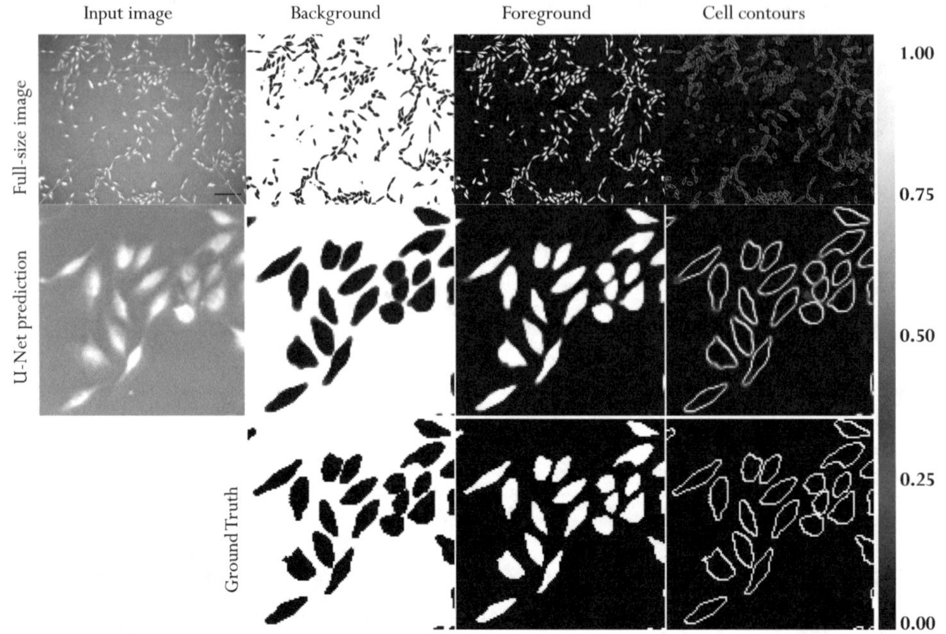

4

◘ Fig. 4.10 Example of network output. Given an input image (top-left, scale bar: 150 μm), the output of our U-Net is an image with three channels, each of them indicating the probability of being background, foreground or cell contour (columns 2–4). The color intensity of the three channels is equally calibrated from 0 to 1. Notice these predictions contain continuous values from 0 to 1 so they need to be post-processed in order to get a binary mask for each class as in the GT (last row). Note that the cells touching the image borders are discarded from the CTC GT

segmentation). So, a distance transform Watershed routine is included in the post-processing macro[20] together with some morphological operations to split cell clusters and refine the results:

```
1    // Rename output image
2    rename("output");
3    // Display in grayscale
4    Stack.setDisplayMode("grayscale");
5    // pseudo-"argmax" operation (from one-hot encoding to 0-1-2 labels)
6    setThreshold(0.5, 1.0);
7    setOption("BlackBackground", true);
8    run("Convert to Mask", "method=Mean background=Dark black");
9    run("Divide...", "value=255.000 stack");
10   setSlice(1);
11   run("Multiply...", "value=0 slice");
12   setSlice(2);
13   run("Multiply...", "value=1 slice");
14   setSlice(3);
15   run("Multiply...", "value=2 slice");
16   run("Z Project...", "projection=[Max Intensity]");
17   rename("argmax");
18   close( "output" )
19   // Analyze foreground (1) label only
20   run("Select Label(s)", "label(s)=1");
21   close("argmax")
22   selectWindow("argmax-keepLabels");
23   // Fill holes
24   run("Fill Holes (Binary/Gray)");
25   close("argmax-keepLabels");
26   // Convert to 0-255
27   run("Multiply...", "value=255.000");
```

20 Macro available at ► https://github.com/NEUBIAS/neubias-springer-book-2021/tree/master/Ch04_Building_a_Bioimage_Analysis_Workflow_using_Deep_Learning/ij-macros/postprocessing.txt.

```
28    // Apply distance transform watershed to extract objects
29    run("Distance Transform Watershed", "distances=[Borgefors (3,4)] output=[32 bits] normalize dynamic=1
↪        connectivity=4");
30    close("argmax-keepLabels-fillHoles");
31    // Remove small objects
32    run("Label Size Filtering", "operation=Greater_Than size=10");
33    close("argmax-keepLabels-fillHoles-dist-watershed");
34    // Rename final image and assign color map
35    selectWindow("argmax-keepLabels-fillHoles-dist-watershed-sizeFilt");
36    rename( "segmented-cells" );
37    run("Set Label Map", "colormap=[Golden angle] background=White shuffle");
38    resetMinAndMax();
```

ImageJ macro for post-processing in DeepImageJ

4.4.6 Step 6: Process All Images in Fiji Using DeepImageJ and MorpholibJ

We are now reaching the final stage of the workflow! We are ready to quantify the morphology of the cells from the test set. Download the data from the CTC web page (▶ Sect. 4.2) and unzip it. Use the Fiji/ImageJ macro provided in this chapter[21] to process the new images. Please, update the path in the macro with the location of the unzipped CTC images in your computer.

With this macro, the individual masks of the cells extracted from the downloaded CTC images will be stored (one label image per input image) together with their corresponding morphological measurements in an easy-to-read comma-separated values (CSV) file (see ◻ Fig. 4.11). More precisely, for each segmented cell, the area, perimeter, circularity, Euler number, bounding box, centroid coordinates, equivalent ellipse, ellipse elongation, convexity, maximum Feret diameter, oriented box, oriented box elongation, geodesic diameter, tortuosity, maximum inscribed disc, average thickness and geodesic elongation will be recorded. For a detailed description of each measurement, see the latest version of MorphoLibJ manual.[22]

Take-Home Message

In this chapter, we have presented a complete bioimage analysis workflow leveraging a DL model to segment cells from phase contrast images. The proposed workflow is versatile and meant to be customizable to other image segmentation-related tasks. As was demonstrated, DL models for bioimage processing can be easily used in Fiji/ImageJ. However, trained models do not perform generally as well on new (and different) images unless they are re-trained. That being said, the proposed workflow can be effortlessly applied to new (similar) datasets by simply modifying the input folders and reproducing the steps described in this document.

---

21 Macro available at ▶ https://github.com/NEUBIAS/neubias-springer-book-2021/blob/master/Ch04_Building_a_Bioimage_Analysis_Workflow_using_Deep_Learning/ij-macros/Step-5-process-folder.ijm.

22 ▶ https://github.com/ijpb/MorphoLibJ/releases/download/v1.4.3/MorphoLibJ-manual-v1.4.3.pdf.

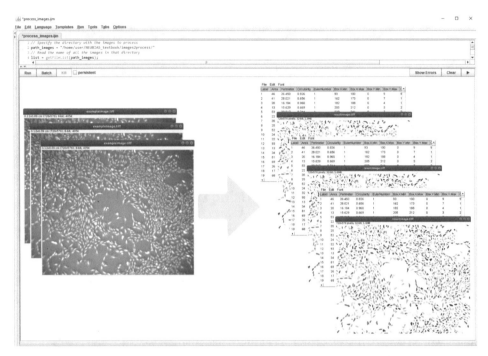

□ Fig. 4.11 Final step. From an ImageJ macro, the images stored in the folder *images_to_process* are processed using the trained model and for each detected cell, a complete list of morphological features are calculated

Acknowledgements   This work is partially supported by the Spanish Ministry of Economy and Competitiveness (TEC2015-73064-EXP, TEC2016-78052-R and PID2019-109820RB-I00) and by a 2017 and a 2020 Leonardo Grant for Researchers and Cultural Creators, BBVA Foundation. We thank the program "Short Term Scientific Missions" of NEUBIAS (network of European bioimage analysists). We also want to acknowledge the support of NVIDIA Corporation with the donation of the Titan X (Pascal) GPU board used for this research. We would like to thank the continuous support of DeepImageJ contributors: C. García-López-de-Haro (UC3M) and D. Sage (EPFL).

Appendix

Training Hyper-Parameters

The hyper-parameters of a DL model (e.g., number of filters) affect the training process and the final result or instance of the model. The study of how to optimize hyper-parameters is a field itself in Computer Vision. Note that the training is a stochastic procedure for which it is almost impossible to reproduce exactly the same training schedule. Additionally, optimizing the combination of hyper-parameters is an exhausting task due to the large amount of time and complexity it requires. In the following paragraphs, we provide you with some tips on how to adjust the most common hyper-parameters:

- **Size of the convolution kernels:** The larger the kernel size is, the wider the receptive field of the CNN is. Namely, the size of the region in the input that produces the feature is larger. However, it is unusual to see kernel sizes larger than 5×5 as it compromises the use of memory. Note that 3D convolutions are also available in Keras (Conv3D) and are defined using 4 dimensions: filters and size of the kernel.
- **Number of features in the convolutional layers:** The more layers and features in a network, the deeper it is and, in theory, the higher its generalizing capacity. For the U-Net, it is recommended to start at least with 16 features in the first convolutional layer and double it as the encoder path becomes deeper (◘ Fig. 4.7).
- **Learning rate:** We choose 0.0003 since experimentally, it was the value for which we got best results. Nevertheless, we tried other values such as 0.001, 0.0005 and 0.0001, as the choice of an optimal learning rate value still remains a trial-and-error problem.
- **Number of training epochs:** It is recommended to set a high value, monitor the training and stop it once you are satisfied with the result (high accuracy, no over-fitting) or you see no improvement.

Optimizer

There are three most common optimizers (SGD, Adam and RMSprop). We chose Adam empirically as, for this dataset, it makes the training improving faster. Adam is also a very common choice, since it is a computationally efficient optimizer that adapts the learning rate to produce smooth convergence. SGD, on the other hand, maintains a single learning rate value for all weight updates during the whole training and can therefore get stuck in local minimum. That being said, and although SGD takes more time to train the model, it sometimes leads to a better generalization of the network.

Halo and Receptive Field of a Network

The output of a single convolution has a smaller size than the input image, unless extra values are added around the image (i.e., padding is performed). For CNN, the pixels in the contours of the output image need to be discarded. The halo is equal to the cumulative padding performed along the CNN and it is determined by the receptive field of one pixel (R) in the U-Net:

$$R = 2^p \left(l \left(\frac{k-1}{2} \right) \right) + 2 \sum_{i=0}^{p-1} 2^i \left(l \left(\frac{k-1}{2} \right) \right) \tag{4.3}$$

where k is the kernel size for each convolutional layer, p is the number of poolings, and l is the number of convolutional layers at each level of the U-Net. In our case

$k = 3$, $p = 3$ and $l = 2$, so the receptive field is 44 (see ◻ Fig. 4.7). In Eq. 4.3, it is assumed that the encoding and decoding paths are symmetrical (i.e. same number of down and up-samplings). Likewise, it is assumed that all convolutional kernels are squared and are all of the same size. The last one, as it is of size 1×1, does not affect the final result of R. See Appendix in the notebook for a computational solution when the analytical expression for R is not available.

Data Augmentation

Increasing the amount of data when training a DL model can improve its capacity to generalize and its performance. However, and often in the biomedical field, obtaining annotated data is difficult and expensive. Therefore, a common technique called data augmentation (DA) is used in DL to provide the model with more unseen data. It consists on creating *new* images applying some transformations to the original ones (i.e., flips, shearing, shifting, zooming). More complex techniques such as elastic transformations, contrast changes or blurring can also be used.[23] The goal is to generate plausible images, so not all the transformations may necessarily improve the training process. For instance, applying contrast variations in the DA process may hinder the learning process if those are not present in the real image data set.

Here we present a common DA implementation based on Keras class `ImageDataGenerator` and its inner `flow()` function, that allows us to choose between a bunch of different transformations.[24] Its implementation enables DA on the fly: it applies a random transformation to each image patch before feeding it to the network. Hence, in each iteration, a *new* sample not seen before is used to train the network. Note that the channels of each mask should be transformed together with their corresponding image. This can be ensured by (1) choosing the same generator configuration for each of the channels in the masks and the input images (`X_train`), and (2) setting the `seed` parameter to the same value for all the cases.

The following code contains a function that creates a DA generator to transform the image patches. By default, the applied transforms include a random choice between 90, 180 or 270 rotations, and vertical and horizontal flips:

```
1   from tensorflow.keras.preprocessing.image import ImageDataGenerator
2   from skimage import transform
3   def join_generators( x_gen, y_gen1, y_gen2, y_gen3 ):
4       while True:
5           x = x_gen.next()
6           y1 = y_gen1.next()
7           y2 = y_gen2.next()
8           y3 = y_gen3.next()
9           yield x, np.concatenate( (y1, y2, y3), axis=-1 )
10  # Random rotation of an image by a multiple of 90 degrees
11  def random_90rotation( img ):
12      return transform.rotate(img, 90*np.random.randint( 0, 5 ), preserve_range=True)
13  # Runtime data augmentation
14  def get_train_val_generators(X_train, Y_train, X_val, Y_val,
15                               batch_size=32, seed=42, rotation_range=0,
16                               horizontal_flip=True, vertical_flip=True,
17                               width_shift_range=0.0,
18                               height_shift_range=0.0,
19                               shear_range=0.0,
20                               brightness_range=None,
21                               rescale=None,
```

23 A good python library to implement DA generators with a wide variety of transformations: ▶ https://github.com/aleju/imgaug.

24 Notebook with the implementation: ▶ https://github.com/NEUBIAS/neubias-springer-book-2021/blob/master/Ch04_Building_a_Bioimage_Analysis_Workflow_using_Deep_Learning/notebook/U_Net_PhC_C2DL_PSC_segmentation_DA.ipynb.

```
22                           preprocessing_function=None,
23                           show_examples=False):
24      # Image data generator distortion options
25      data_gen_args = dict( rotation_range = rotation_range,
26                            width_shift_range=width_shift_range,
27                            height_shift_range=height_shift_range,
28                            shear_range=shear_range,
29                            brightness_range=brightness_range,
30                            preprocessing_function=preprocessing_function,
31                            horizontal_flip=horizontal_flip,
32                            vertical_flip=vertical_flip,
33                            fill_mode='reflect')
34
35      # Train data, provide the same seed and keyword arguments to the fit and flow methods
36      # (one datagen per class)
37      Y_datagen1 = ImageDataGenerator(**data_gen_args)
38      Y_datagen2 = ImageDataGenerator(**data_gen_args)
39      Y_datagen3 = ImageDataGenerator(**data_gen_args)
40
41      Y_train1 = np.expand_dims( Y_train[:,:,:,0], axis=-1 )
42      Y_train2 = np.expand_dims( Y_train[:,:,:,1], axis=-1 )
43      Y_train3 = np.expand_dims( Y_train[:,:,:,2], axis=-1 )
44
45      data_gen_args['rescale'] = rescale # rescale only X, not Y
46      X_datagen = ImageDataGenerator(**data_gen_args)
47
48      X_datagen.fit(X_train, augment=True, seed=seed)
49      Y_datagen1.fit(Y_train1, augment=True, seed=seed)
50      Y_datagen2.fit(Y_train2, augment=True, seed=seed)
51      Y_datagen3.fit(Y_train3, augment=True, seed=seed)
52
53      X_train_augmented = X_datagen.flow(X_train, batch_size=batch_size, shuffle=True, seed=seed)
54      Y_train_augmented1 = Y_datagen1.flow(Y_train1, batch_size=batch_size, shuffle=True, seed=seed)
55      Y_train_augmented2 = Y_datagen2.flow(Y_train2, batch_size=batch_size, shuffle=True, seed=seed)
56      Y_train_augmented3 = Y_datagen3.flow(Y_train3, batch_size=batch_size, shuffle=True, seed=seed)
57
58      # Validation data, no data augmentation, but we create a generator anyway
59      X_datagen_val = ImageDataGenerator(rescale=rescale)
60      Y_datagen_val1 = ImageDataGenerator()
61      Y_datagen_val2 = ImageDataGenerator()
62      Y_datagen_val3 = ImageDataGenerator()
63
64      Y_val1 = np.expand_dims( Y_val[:,:,:,0], axis=-1 )
65      Y_val2 = np.expand_dims( Y_val[:,:,:,1], axis=-1 )
66      Y_val3 = np.expand_dims( Y_val[:,:,:,2], axis=-1 )
67
68      X_datagen_val.fit(X_val, augment=True, seed=seed)
69      Y_datagen_val1.fit(Y_val1, augment=True, seed=seed)
70      Y_datagen_val2.fit(Y_val2, augment=True, seed=seed)
71      Y_datagen_val3.fit(Y_val3, augment=True, seed=seed)
72
73      X_val_augmented = X_datagen_val.flow(X_val, batch_size=batch_size, shuffle=False, seed=seed)
74      Y_val_augmented1 = Y_datagen_val1.flow(Y_val1, batch_size=batch_size, shuffle=False, seed=seed)
75      Y_val_augmented2 = Y_datagen_val2.flow(Y_val2, batch_size=batch_size, shuffle=False, seed=seed)
76      Y_val_augmented3 = Y_datagen_val3.flow(Y_val3, batch_size=batch_size, shuffle=False, seed=seed)
77      if show_examples:
78          plt.figure(figsize=(20,15))
79          # Column titles
80          cols = ['Original', 'Augmented', 'Augmented Binary Mask', 'Augmented Binary Mask', 'Augmented Contour Mask']
81          # Create a augmentor just to show original images together with samples
82          X_train_original = X_datagen_val.flow(X_train, batch_size=batch_size, shuffle=True, seed=seed)
83          # generate samples and plot
84          for i in range(3):
85              # Original image plot
86              ax = plt.subplot(3,5,1 + 5*i)
87              ax.title.set_text(cols[0])
88              batch = X_train_original.next()
89              image = batch[0]
90              plt.imshow(image[:,:,0], vmin=0, vmax=1, cmap='gray')
91              # Augmented image
92              ax = plt.subplot(3,5,1 + 5*i+1)
93              ax.title.set_text(cols[1])
94              batch = X_train_augmented.next()
95              image = batch[0]
96              plt.imshow(image[:,:,0], vmin=0, vmax=1, cmap='gray')
97              # Augmented Binary Mask (Background)
98              ax = plt.subplot(3,5,1 + 5*i+2)
99              ax.title.set_text(cols[2])
00              batch = Y_train_augmented1.next()
01              image = batch[0]
02              plt.imshow(image[:,:,0],  cmap='gray', interpolation='nearest' )
03              # Augmented Binary Mask (Cells)
04              ax = plt.subplot(3,5,1 + 5*i+3)
05              ax.title.set_text(cols[3])
06              batch = Y_train_augmented2.next()
07              image = batch[0]
08              plt.imshow(image[:,:,0],  cmap='gray', interpolation='nearest' )
09              #Augmented Contour Mask
10              ax = plt.subplot(3,5,1 + 5*i+4)
11              ax.title.set_text(cols[4])
12              batch = Y_train_augmented3.next()
13              image = batch[0]
14              plt.imshow(image[:,:,0],  cmap='gray', interpolation='nearest' )
15          plt.show()
16          del X_train_original
```

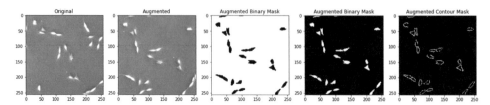

Fig. 4.12 Output of previous code displaying an original image and the transformation made by the generators. The binary masks have been transformed in the same way

```
117         X_train_augmented.reset()
118         Y_train_augmented1.reset()
119         Y_train_augmented2.reset()
120         Y_train_augmented3.reset()
121
122         # combine generators into one which yields image and masks
123         n_train = X_train_augmented.n
124         train_generator = join_generators( X_train_augmented, Y_train_augmented1,
125                                            Y_train_augmented2, Y_train_augmented3 )
126         n_val = X_val_augmented.n
127         val_generator = join_generators( X_val_augmented, Y_val_augmented1,
128                                          Y_val_augmented2, Y_val_augmented3 )
129         return train_generator, val_generator, n_train, n_val
```

code cell[11], U_Net_PhC_C2DL_PSC_segmentation_DA.ipynb

As the network is fed with generators, a validation data generator needs to be created. Note that we create it without applying any transformation to the data, as it needs to be unchanged and always the same to ensure a correct validation of the model.

The following code calls the previous function to create the data generators and displays a few images ensuring that the generators produce a transformed version of the original ones together with their associated masks (see ◻ Fig. 4.12):

```
1  train_generator, val_generator, \
2      n_train, n_val = get_train_val_generators(X_train=X_train,
3                                                Y_train=Y_train,
4                                                X_val=X_val,
5                                                Y_val=Y_val,
6                                                rescale=None,
7                                                horizontal_flip=True,
8                                                vertical_flip=True,
9                                                shear_range=0,
10                                               zoom_range=0,
11                                               rotation_range = 0,
12                                               batch_size=batchSize,
13                                               preprocessing_function=random_90rotation,
14                                               show_examples=True)
```

code cell [14], U_Net_PhC_C2DL_PSC_segmentation_DA.ipynb

As we are training with generators, on this version of Tensorflow the function to train the network must be changed to `fit_generator()` instead of `fit()`. Thus, the following code should be used:

```
1  history = model.fit_generator(train_generator, validation_data=val_generator,
2                                steps_per_epoch=int(n_train/batchSize),
3                                validation_steps=int(n_val/batchSize),
4                                epochs=numEpochs, callbacks=[earlystopper])
```

code cell [15], U_Net_PhC_C2DL_PSC_segmentation_DA.ipynb

Solutions to the Exercises

✅ Exercise 1

```
1    # Path to the validation images
2    val_input_path = '/content/dataset/validation_input'
3    val_masks_path ='/content/dataset/validation_binary_masks'
4    val_contours_path = '/content/dataset/validation_contours'
5    # Read the list of file names and sort them to have a match between images and masks
6    val_input_filenames = [x for x in os.listdir(val_input_path ) if x.endswith(".tif")]
7    val_input_filenames.sort()
8    val_masks_filenames = [x for x in os.listdir(val_masks_path ) if x.endswith(".tif")]
9    val_masks_filenames.sort()
10   val_contours_filenames = [x for x in os.listdir(val_contours_path ) if x.endswith(".png")]
11   val_contours_filenames.sort()
12   # read training images
13   val_img = [cv2.imread(os.path.join(val_input_path, x), cv2.IMREAD_ANYDEPTH) for x in val_input_filenames ]
14   val_masks = [cv2.imread(os.path.join(val_masks_path, x), cv2.IMREAD_ANYDEPTH)>0 for x in val_masks_filenames ]
15   val_contours = [cv2.imread(os.path.join(val_contours_path, x), cv2.IMREAD_ANYDEPTH)>0 for x in
↪    val_contours_filenames ]
16   # concatenate binary masks and contours
17   val_output = [np.transpose(np.array([val_masks[i],val_contours[i]]), [1,2,0]) for i in range(len(val_masks))]
18   # Create the validation patches
19   X_val, val_output_patches = create_random_patches( val_img, val_output, 6, [256,256])
20   # In Y_val we will store the one-hot respresentation of the labels
21   Y_val = [np.stack([1 - x[:,:,0] - x[:,:,1], x[:,:,0], x[:,:,1]], axis=-1) for x in val_output_patches ]
22   Y_val = np.asarray(Y_val)
23   # In X_val we store the input patches of the validation set
24   X_val = [np.expand_dims(x, axis=-1) for x in X_val]
25   X_val = np.asarray(X_val)
```

code cell [6], U_Net_PhC_C2DL_PSC_segmentation.ipynb

✅ Exercise 2

Add the following lines to the code in Step 3.1, right after reading the files in the training data directory. It will reduce the training data set to 10 images:

```
1    # Create 10 random numbers to reduce the training data set:
2    import numpy as np
3    index = np.random.randint(len(train_input_filenames)-1, size=10, dtype=np.int)
4    # Reduce the training set
5    train_input_filenames = [train_input_filenames[i] for i in index]
6    train_masks_filenames = [train_masks_filenames[i] for i in index]
7    train_contours_filenames = [train_contours_filenames[i] for i in index]
8    print( 'Number of training input images: ' + str( len(train_input_filenames)) )
9    print( 'Number of training binary mask images: ' + str( len(train_masks_filenames)) )
10   print( 'Number of training contour images: ' + str( len(train_contours_filenames)) )
```

Alternative code cell [3], U_Net_PhC_C2DL_PSC_segmentation.ipynb

For this example, we set the number of epochs to 1000 in the Step 3.4, and run the entire code to train the network from scratch using 10 images. The result is as follows:

```
1    Train on 60 samples, validate on 66 samples
2    Epoch 1/1000
3    60/60 [==============================] - 13s 214ms/step - loss: 0.5979 - jaccard_index: 0.0075 - val_loss:
↪    0.4125 - val_jaccard_index: 0.0000e+00
4    ...
5    Epoch 418/1000
6    60/60 [==============================] - 2s 32ms/step - loss: 0.0377 - jaccard_index: 0.8101 - val_loss: 0.0789
↪    - val_jaccard_index: 0.7415
7    Restoring model weights from the end of the best epoch
8    Epoch 00418: early stopping
9    # Evaluation of the model in the test set:
10   test loss CCE: 0.12604248134626284, Jaccard index: 0.705814957143211
```

Results of Exercise 2

✅ Exercise 3

```
1    # Now we load some unseen images for testing
2    test_input_path = '/content/dataset/test_input'
3    test_masks_path ='/content/dataset/test_binary_masks'
4    test_contours_path = '/content/dataset/test_contours'
5    test_input_filenames = [x for x in os.listdir( test_input_path ) if x.endswith(".tif")]
6    test_input_filenames.sort()
7    test_mask_filenames = [x for x in os.listdir( test_masks_path ) if x.endswith(".tif")]
8    test_mask_filenames.sort()
9    test_contours_filenames = [x for x in os.listdir(test_contours_path ) if x.endswith(".png")]
```

```
10   test_contours_filenames.sort()
11   # Read test images
12   test_img = [cv2.imread(os.path.join(test_input_path, x), cv2.IMREAD_ANYDEPTH) for x in test_input_filenames ]
13   test_masks = [cv2.imread(os.path.join(test_masks_path, x), cv2.IMREAD_ANYDEPTH)>0 for x in test_mask_filenames ]
14   test_contours = [cv2.imread(os.path.join(test_contours_path, x), cv2.IMREAD_ANYDEPTH)>0 for x in
     ↪  test_contours_filenames ]
15   # concatenate binary masks and contours
16   test_output = [np.transpose(np.array([test_masks[i],test_contours[i]]), [1,2,0]) for i in
     ↪  range(len(test_masks))]
17   # Adapt the test images to an appropriate size using the same function as before
18   test_input_patches, test_output_patches = create_random_patches( test_img, test_output, 1, [560,704])
19   # Normalize input imagess
20   X_test = [x/255 for x in test_input_patches] # normalize between 0 and 1
21   X_test = [np.expand_dims(x, axis=-1) for x in X_test]
22   X_test = np.asarray(X_test)
23   # One-hot label representation
24   Y_test = [np.stack([1 - x[:,:,0] - x[:,:,1], x[:,:,0], x[:,:,1]], axis=-1) for x in test_output_patches ]
25   Y_test = np.asarray(Y_test)
```

code cell [14-15], U_Net_PhC_C2DL_PSC_segmentation.ipynb

References

Abadi M, Barham P, Chen J, Chen Z, Davis A, Dean J, Devin M, Ghemawat S, Irving G, Isard M, et al. (2016) Tensorflow: A system for large-scale machine learning. In: 12th {USENIX} symposium on operating systems design and implementation ({OSDI} 16), p 265–283

Bankhead P, Loughrey MB, Fernández JA, Dombrowski Y, McArt DG, Dunne PD, McQuaid S, Gray RT, Murray LJ, Coleman HG et al (2017) Qupath: open source software for digital pathology image analysis. Sci Rep 7(1):1–7

Berg S, Kutra D, Kroeger T, Straehle CN, Kausler BX, Haubold C, Schiegg M, Ales J, Beier T, Rudy M, Eren K, Cervantes JI, Xu B, Beuttenmueller F, Wolny A, Zhang C, Koethe U, Hamprecht FA, Kreshuk A (2019) ilastik: interactive machine learning for (bio)image analysis. Nat Methods 16:1226–1232. https://doi.org/10.1038/s41592-019-0582-9

Bisong E (2019) Google colaboratory. Building machine learning and deep learning models on google cloud platform. Springer, Berlin, pp 59–64

Cardona A, Saalfeld S, Schindelin J, Arganda-Carreras I, Preibisch S, Longair M, Tomancak P, Hartenstein V, Douglas RJ (2012) Trakem2 software for neural circuit reconstruction. PLoS One 7(6):e38011

Chollet F, et al. (2015) keras

Falk T, Mai D, Bensch R, Çiçek Ö, Abdulkadir A, Marrakchi Y, Böhm A, Deubner J, Jäckel Z, Seiwald K et al (2019) U-net: deep learning for cell counting, detection, and morphometry. Nat Methods 16(1):67–70

Goodfellow I, Bengio Y, Courville A (2016) Deep learning. MIT Press, Cambridge. http://www.deeplearningbook.org

Kapur T, Pieper S, Fedorov A, Fillion-Robin JC, Halle M, O'Donnell L, Lasso A, Ungi T, Pinter C, Finet J et al (2016) Increasing the impact of medical image computing using community-based open-access hackathons: the NA-MIC and 3d slicer experience. Med Image Anal 33:176–180

Kiefer J, Wolfowitz J et al (1952) Stochastic estimation of the maximum of a regression function. Ann Math Stat 23(3):462–466

Kingma DP, Ba J (2014) Adam: a method for stochastic optimization. eprint: 1412.6980

Legland D, Arganda-Carreras I, Andrey P (2016) Morpholibj: integrated library and plugins for mathematical morphology with imagej. Bioinformatics 32(22):3532–3534

Litjens G, Kooi T, Bejnordi BE, Setio AAA, Ciompi F, Ghafoorian M, Van Der Laak JA, Van Ginneken B, Sánchez CI (2017) A survey on deep learning in medical image analysis. Med Image Anal 42:60–88

Gómez-de Mariscal E, García-López-de Haro C, Donati L, Unser M, Muñoz-Barrutia A, Sage D (2019) Deepimagej: a user-friendly plugin to run deep learning models in imagej. bioRxiv p 799270

Maška M, Ulman V, Svoboda D, Matula P, Matula P, Ederra C, Urbiola A, España T, Venkatesan S, Balak DM et al (2014) A benchmark for comparison of cell tracking algorithms. Bioinformatics 30(11):1609–1617

Ouyang W, Mueller F, Hjelmare M, Lundberg E, Zimmer C (2019) Imjoy: an open-source computational platform for the deep learning era. Nat Methods 16(12):1199–1200

Paszke A, Gross S, Massa F, Lerer A, Bradbury J, Chanan G, Killeen T, Lin Z, Gimelshein N, Antiga L, Desmaison A, Kopf A, Yang E, DeVito Z, Raison M, Tejani A, Chilamkurthy S, Steiner B, Fang

L, Bai J, Chintala S (2019) Pytorch: An imperative style, high-performance deep learning library. In: Wallach H, Larochelle H, Beygelzimer A, d'Alché-Buc F, Fox E, Garnett R (eds) Advances in neural information processing systems, vol. 32. Curran Associates, Red Hook, p 8024–8035. http://papers.neurips.cc/paper/9015-pytorch-an-imperative-style-high-performance-deep-learning-library.pdf

Patterson J, Gibson A (2017) Deep learning: a practitioner's approach. O'Reilly, Beijing. https://www.safaribooksonline.com/library/view/deep-learning/9781491924570/

Roh Y, Heo G, Whang SE (2021) A survey on data collection for machine learning: a big data-AI integration perspective. IEEE Trans Knowl Data Eng 33(4):1328–1347

Ronneberger O, Fischer P, Brox T (2015) U-net: Convolutional networks for biomedical image segmentation. International conference on medical image computing and computer-assisted intervention. Springer, Berlin, pp 234–241

Rosenblatt F (1961) Principles of neurodynamics. perceptrons and the theory of brain mechanisms. Tech. rep., Cornell Aeronautical Lab Inc Buffalo NY

Rueden CT, Schindelin J, Hiner MC, DeZonia BE, Walter AE, Arena ET, Eliceiri KW (2017) Imagej 2: Imagej for the next generation of scientific image data. BMC Bioinf 18(1):529

Saalfeld S, Cardona A, Hartenstein V, Tomančák P (2009) CATMAID: collaborative annotation toolkit for massive amounts of image data. Bioinformatics 25(15):1984–1986. https://doi.org/10.1093/bioinformatics/btp266. https://academic.oup.com/bioinformatics/article-pdf/25/15/1984/555362/btp266.pdf

Schindelin J, Arganda-Carreras I, Frise E, Kaynig V, Longair M, Pietzsch T, Preibisch S, Rueden C, Saalfeld S, Schmid B et al (2012) Fiji: an open-source platform for biological-image analysis. Nat Methods 9(7):676–682

Schneider CA, Rasband WS, Eliceiri KW (2012) Nih image to imagej: 25 years of image analysis. Nat Methods 9(7):671–675

Sofroniew N, Lambert T, Evans K, Nunez-Iglesias J, Yamauchi K, Solak AC, Bokota G, ziyangczi, Buckley G, Winston P, Tung T, Pop DD, Hector, Freeman J, Bussonnier M, Boone P, Royer L, Har-Gil H, Axelrod S, Rokem A, Bryant, Kiggins J, Huang M, Vemuri P, Dunham R, Manz T, jakirkham, Wood C, de Siqueira A, Chopra B (2020) napari/napari: 0.3.8rc2. https://doi.org/10.5281/zenodo.4048613

Ulman V, Maška M, Magnusson KE, Ronneberger O, Haubold C, Harder N, Matula P, Matula P, Svoboda D, Radojevic M et al (2017) An objective comparison of cell-tracking algorithms. Nat Methods 14(12):1141–1152

von Chamier L, Jukkala J, Spahn C, Lerche M, Hernández-Pérez S, Mattila PK, Karinou E, Holden S, Solak AC, Krull A, Buchholz TO, Jug F, Royer LA, Heilemann M, Laine RF, Jacquemet G, Henriques R (2020) Zerocostdl4mic: an open platform to simplify access and use of deep-learning in microscopy. https://doi.org/10.1101/2020.03.20.000133. https://www.biorxiv.org/content/early/2020/03/20/2020.03.20.000133, https://www.biorxiv.org/content/early/2020/03/20/2020.03.20.000133.full.pdf

Weigert M, Schmidt U, Boothe T, Müller A, Dibrov A, Jain A, Wilhelm B, Schmidt D, Broaddus C, Culley S, Rocha-Martins M, Segovia-Miranda F, Norden C, Henriques R, Zerial M, Solimena M, Rink J, Tomancak P, Royer L, Jug F, Myers EW (2018) Content-aware image restoration: pushing the limits of fluorescence microscopy. Nat Methods 15(12):1090–1097. https://doi.org/10.1038/s41592-018-0216-7

Yushkevich PA, Piven J, Cody Hazlett H, Gimpel Smith R, Ho S, Gee JC, Gerig G (2006) User-guided 3D active contour segmentation of anatomical structures: significantly improved efficiency and reliability. Neuroimage 31(3):1116–1128

GPU-Accelerating ImageJ Macro Image Processing Workflows Using CLIJ

Daniela Vorkel and Robert Haase

Contents

This Chapter has been reviewed by Dominic Waithe, University of Oxford.

© The Author(s) 2022
K. Miura, N. Sladoje (eds.), *Bioimage Data Analysis Workflows–Advanced Components and Methods*,
Learning Materials in Biosciences, https://doi.org/10.1007/978-3-030-76394-7_5

What You Will Learn in This Chapter

This chapter introduces GPU-accelerated image processing in ImageJ/Fiji. The reader is expected to have some pre-existing knowledge of ImageJ Macro programming. Core concepts such as variables, *for*-loops, and functions are essential. The chapter provides basic guidelines for improved performance in typical image processing workflows. We present in a step-by-step tutorial how to translate a pre-existing ImageJ macro into a GPU-accelerated macro.[1]

5.1 Introduction

5

Modern life science increasingly relies on microscopic imaging followed by quantitative bioimage analysis (BIA). Nowadays, image data scientists join forces with artificial intelligence researchers, incorporating more and more machine learning algorithms into BIA workflows. Even though general machine learning and convolutional neural networks are not new approaches to image processing, their importance for life science is increasing.

As their application is now at hand due to the rise of advanced computing hardware, namely graphics processing units (GPUs), a natural question is if GPUs can also be exploited for classic image processing in ImageJ (Schneider et al., 2012) and Fiji (Schindelin et al., 2012). As an alternative to established acceleration techniques, such as ImageJ's batch mode, we explore how GPUs can be exploited to accelerate classic image processing. Our approach, called CLIJ (Haase et al., 2020), enables biologists and bioimage analysts to speed up time-consuming analysis tasks by adding support for the Open Computing Language (OpenCL) for programming GPUs (Khronos-Group, 2020) in ImageJ. We present a guide for transforming state-of-the-art image processing workflows into GPU-accelerated workflows using the ImageJ Macro language. Our suggested approach neither requires a profound expertise in high performance computing, nor to learn a new programming language such as OpenCL.

To demonstrate the procedure, we translate a formerly published BIA workflow for examining signal intensity changes at the nuclear envelope, caused by cytoplasmic redistribution of a fluorescent protein (Miura, 2020). We then introduce ways to discover CLIJ commands as counterparts of classic ImageJ methods. These commands are then assembled to refactor the pre-existing workflow. In terms of image processing, refactoring means restructuring an existing macro without changing measurement results, but rather improving processing speed. Accordingly, we show how to measure workflow performance. We also give an insight into quality assurance methods, which help to ensure good scientific practice when modernizing BIA workflows and refactoring code.

---

1   This chapter was communicated by Dominic Waithe, University of Oxford, UK.

5.2 The Dataset

5.2.1 Imaging Data

Cell membranes create functional compartments and maintain diverse content and activities. Fluorescent labeling techniques allow the study of certain structures and cell components, in particular to trace dynamic processes over time, such as changes in intensity and spatial distribution of fluorescent signals. The method of live-cell imaging, taken as long-term time-lapses, is important when studying dynamic biological processes. As a representative dataset for this domain, we process a two-channel time-lapse showing a HeLa cell with increasing signal intensity in one channel (Boni et al., 2015). The dataset has a pixel size of $0.165\,\mu m$ per pixel and a frame delay of $400\,s$. The nuclei-channel (C1), excited with 561 nm wavelength light, consists of Histone H2B-mCherry signals within the nucleus. The protein-channel (C2), excited with 488 nm wavelength light, represents the distribution of the cytoplasmic Lamin B protein, which accumulates at the inner nuclear membrane (Lamin B receptor signal). Four example time points of the dataset are shown in ◘ Fig. 5.1.

5.2.2 The Predefined Processing Workflow

To measure the changing intensities along the nuclear envelope, it is required to define a corresponding region of interest (ROI) within the image. First, the image is segmented into nucleus and background. Second, a region surrounding the nucleus is derived.

A starting point for the workflow translation is the *code_final.ijm* macro file published by Miura (2020).[2] For reader's convenience, we have added some explanatory comments for each section of the original code:

```
1  // determine current data set and split channels
2  orgName = getTitle();
3  run("Split Channels");
4  c1name = "C1-" + orgName;
5  c2name = "C2-" + orgName;
```

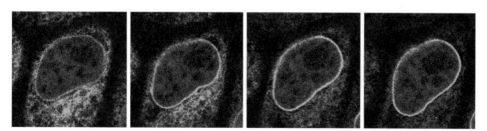

◘ **Fig. 5.1**   Samples of the dataset used in this chapter: Time points 1, 5, 10 and 15, showing the signal increase in the nuclear envelope of a cell. Courtesy: Andrea Boni, EMBL Heidelberg/Viventis

---

2   ▶ https://github.com/miura/NucleusRimIntensityMeasurementsV2/blob/master/code/code_final. ijm.

5

```
6
7    // invoke segmentation of a band around the nucleus
8    selectWindow(c1name);
9    nucorgID = getImageID();
10   nucrimID = nucseg( nucorgID );
11
12   // go through all time points and measure intensity in the band
13   selectWindow(c2name);
14   c2id = getImageID();
15   opt = "area mean centroid perimeter shape integrated display
      ↪  redirect=None decimal=3";
16   run("Set Measurements...", opt);
17   for (i =0; i < nSlices; i++){
18       selectImage( nucrimID );
19       setSlice( i + 1 );
20       run("Create Selection");
21       run("Make Inverse");
22       selectImage( c2id );
23       setSlice( i + 1 );
24       run("Restore Selection");
25       run("Measure");
26   }
27
28   // detailed segmentation of the band around the nucleus
29   function nucseg( orgID ){
30       selectImage( orgId );
31       run("Gaussian Blur...", "sigma=1.50 stack");
32       setAutoThreshold("Otsu dark");
33       setOption("BlackBackground", true);
34       run("Convert to Mask", "method=Otsu background=Dark calculate
          ↪  black");
35       run("Analyze Particles...", "size=800-Infinity pixel
          ↪  circularity=0.00-1.00 show=Masks display exclude clear
          ↪  include stack");
36       dilateID = getImageID();
37       run("Invert LUT");
38       options =  "title = dup.tif duplicate range=1-" + nSlices;
39       run("Duplicate...", options);
40       erodeID = getImageID();
41       selectImage(dilateID);
42       run("Options...", "iterations=2 count=1 black edm=Overwrite
          ↪  do=Nothing");
43       run("Dilate", "stack");
44       selectImage(erodeID);
45       run("Erode", "stack");
46       imageCalculator("Difference create stack", dilateID, erodeID);
47       resultID = getImageID();
48       selectImage(dilateID);
49       close();
50       selectImage(erodeID);
51       close();
52       selectImage(orgID);
53       close();
54       run("Clear Results");
55       return resultID;
56   }
```

5.3 Tools: CLIJ

Available as optional plugin, CLIJ brings GPU-accelerated image processing routines to Fiji. Installation of CLIJ is done by using the Fiji updater, which can be found in the menu *Help > Update*, and by activating the update sites of clij and clij2, as shown in ◘ Fig. 5.2. Depending on GPU vendor and operating system, further installation of GPU drivers might be necessary. In some cases, default drivers delivered by automated operating system updates are not sufficient.

    After installing CLIJ, it is recommended to execute a CLIJ macro to test for successful installation. We can also use this opportunity to get a first clue about a potential speedup of a CLIJ method compared to its ImageJ counterpart. The following example macro processes an image using both methods, and writes the processing time into the log window, as shown in ◘ Fig. 5.3.

```
// load example dataset
run("T1 Head (2.4M, 16-bits)");

// initialize GPU
run("CLIJ2 Macro Extensions", "cl_device=");
Ext.CLIJ2_clear();

// apply a mean filter on the GPU
time = getTime();
```

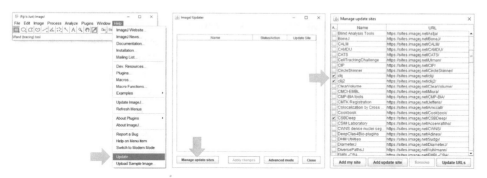

◘ **Fig. 5.2** Installation of CLIJ: In Fiji's updater, which can be found in the menu *Help > Update...*, click on *Manage Update Sites*, and activate the checkboxes next to clij and clij2. After updating and restarting Fiji, CLIJ is installed

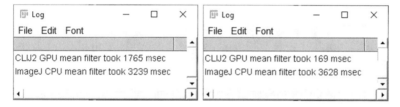

◘ **Fig. 5.3** Output of the first example macro, which reports processing time of a CLIJ operation (first line), and of the classic ImageJ operation (second line). When executing a second time (right), the GPU typically becomes faster due to the so-called warm-up effect

```
input = getTitle();
Ext.CLIJ2_push(input);
Ext.CLIJ2_mean3DBox(input, result, 3, 3, 3);
Ext.CLIJ2_pull(result);
Ext.CLIJ2_clear();
print("CLIJ2 GPU mean filter took " + (getTime() - time) + " msec");

// apply the corresponding operation of classic ImageJ
time = getTime();
run("Mean 3D...", "x=3 y=3 z=3");
print("ImageJ CPU mean filter took " + (getTime() - time) + " msec");
```

5.3.1 Basics of GPU-Accelerated Image Processing with CLIJ

Every ImageJ macro, which uses CLIJ functionality, needs to contain some additional code sections. For example, this is how the GPU is initialized:

```
run("CLIJ2 Macro Extensions", "cl_device=");
Ext.CLIJ2_clear();
```

In the first line, the parameter *cl_device* can stay blank, imposing that CLIJ will select automatically an OpenCL device, namely the GPU. One can specify the name of the GPU in brackets, for example *nVendor Awesome Intelligent*. If only a part of the name is specified, such as *nVendor* or *some*, CLIJ will select a GPU which contains that part in the name. One can explore available GPU devices by using the menu *Plugins > ImageJ on GPU (CLIJ2) > Macro tools > List available GPU devices*. The second line, in the example shown above, cleans up GPU memory. This command is typically called by the end of a macro. It is not mandatory to write it at the beginning, however, it is recommended while elaborating a new ImageJ macro. A macro under development unintentionally stops every now and then with error messages. Hence, a macro is not executed until the very end, where GPU memory typically gets cleaned up. Thus, it is recommended to write this line initially, to start at a predefined empty state.

Another typical step in CLIJ macros is to push image data to the GPU memory:

```
input = getTitle();
Ext.CLIJ2_push(input);
```

We first retrieve the name of the current image by using ImageJ's built-in *getTitle()*-command, and save it into the variable *input*. Afterwards, the *input* image is stored in GPU memory using CLIJ's push method.

This image can then be processed, for example using a mean filter:

```
Ext.CLIJ2_mean3DBox(input, result, 3, 3, 3);
```

CLIJ's mean filter, applied to a 3D image, takes a cuboidal neighborhood into account, as specified by the word *Box*. It has five parameters: the *input* image name, the *result* image name given by variables, and three half-axis lengths describing the size of the box. If the variable for the result is not set, it will be set to an automatically generated image name.

Finally, the *result*-image gets pulled back from GPU memory and will be displayed on the screen.

```
Ext.CLIJ2_pull(result);
Ext.CLIJ2_clear();
```

Hence, result images are not shown on the screen until the *pull( )* command is explicitly called. Thus, the computer screen is not flooded with numerous image windows, helping the workflow developer to stay organised. Furthermore, memory gets cleaned up by the *clear( )* command, as explained above.

While developing advanced CLIJ workflows, it might be necessary to take a look into GPU memory to figure out which images are stored at a particular moment. Therefore, we can add another command just before the final *clear( )*-command, which will list images in GPU memory in the log windows, as shown in ◘ Fig. 5.4:

```
Ext.CLIJ2_reportMemory();
```

As an intermediate summary, CLIJ commands in ImageJ macro typically appear as follows:

```
Ext.CLIJ2_operation(parameters);
```

All CLIJ methods start with the prefix *Ext.*, a convention by classical ImageJ, indicating that we are calling a macro extension optionally installed to ImageJ. Next, it reads *CLIJ_*, *CLIJ2_* or *CLIJx_* followed by the specific method and, in brackets, the parameters passed over to this method. Parameters are typically given in the order: input images, output images, other parameters.

The *CLIJ* identifier was introduced to classify methods originally published as CLIJ toolbox (Haase et al., 2020). It is now deprecated since the official stable release of *CLIJ2*, which is the extended edition of *CLIJ* including more operations. Furthermore, there is *CLIJx*, the volatile experimental sibling, which is constantly evolving as developers work on it. Thus, CLIJx methods should be used with care as the *X* stands for e*X*perimental. Whenever possible, the latest stable release should be used. As soon as a new stable release is out, the former one will be deprecated. The deprecated release will be kept available for at least 1 year. To allow a convenient transition between major releases, the CLIJ developers strive for backwards-compatibility between releases.

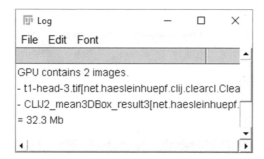

◘ **Fig. 5.4**   List of images currently stored in GPU memory: In this case, there exists an image called *t1-head-3.tif* which corresponds to the dataset we loaded initially. Furthermore, there is another image, called *CLIJ2_mean3DBox_result3*, containing the result of the mean filter operation

5.3.2 Where CLIJ Is Conceptually Different and Why

When designing the CLIJ application programming interface (API), special emphasis was put on a couple of aspects to standardize and simplify image processing.

— Results of CLIJ operations are per default not shown on screen. One needs to pull the image data from the GPU memory to display them in an ImageJ window. In order to achieve optimal performance, it is recommended to execute as many processing steps as possible between push and pull commands. Furthermore, only the final result image should be pulled. Pushing and pulling take time. This time investment can be gained back by calling operations, which are altogether faster than the classic ImageJ operations.

— CLIJ operations always need explicit specifications of input and output images. The currently selected window in ImageJ does not play a role when calling a CLIJ command. Moreover, no command in CLIJ changes the input image. The only exception are commands starting with 'set', which take no input image and overwrite pixels of a given output image. All other commands read pixels from input images and write new pixels into output images, as in the following example:

```
Ext.CLIJ2_excludeLabelsOnEdges(labels,
 ↪   labels_without_touching_edges);
```

— CLIJ operations do not take physical units into account. For example, all radius and sigma parameters are provided in pixel units:

```
sigma = 1.5;
Ext.CLIJ2_gaussianBlur2D(orgID, blurred, sigma, sigma);
```

— If a CLIJ method's name contains the terms "2D" or "3D", it processes, respectively, two- or three-dimensional images. If the name of the method is without such a term, the method processes images of both types.

— Images and image stacks in CLIJ are granular units of data, meaning that individual pixels of an image cannot be accessed efficiently by a GPU. Instead, pixels are processed in parallel, and therefore the whole image at once. Time-lapse data need to be split into image stacks and processed time point by time point.

— CLIJ methods are granular operations on data. That means, they apply a single defined procedure to a whole image. Independent from any ImageJ configuration, CLIJ methods produce the same output given the same input. Conceptually, this leads to improved readability and maintenance of image processing workflows.

5.3.3 Hardware Suitable for CLIJ

When using CLIJ, for best possible performance it is recommended to use recent GPUs. Technically, CLIJ is compatible with GPU-devices supporting the OpenCL 1.2 standard (Khronos-Group, 2020), which was established in 2011. While OpenCL works on GPUs up to 9 years old, GPU devices older than 5 years may be unable to offer a processing performance faster than recent CPUs. Thus, when striving for high performance, recent devices should be utilized. When considering new hardware, image processing specific aspects should be taken into account:

- **Memory size**: State-of-the-art imaging techniques produce granular 2D and 3D image data up to several gigabytes. Dependent on the desired use case, it may make sense to utilize GPUs with increased memory. Convenient workflow development is possible, if a processed image fits about 4–6 times into GPU memory. Hence, if working with images of 1–2 GB in size, a GPU with at least 8 GB of GDDR6 RAM memory should be used.
- **Memory Bandwidth**: Image processing is memory-bound, meaning that all operations have in common that pixels are read from memory and written to memory. Reading and writing is the major bottleneck, and thus, GPUs with fast memory access and with high memory bandwidth should be preferred. Typically, GDDR6-based GPUs have memory bandwidths larger than 400 GB/s. GDDR5-based GPUs often offer less than 100 GB/s. So, GDDR6-based GPUs may compute image processing results about 4 times faster.
- **Integrated GPUs**: For processing of big images, a large amount of memory might be needed. At time of writing, GDDR6-based GPUs with 8 GB of memory are available in price ranges between 300 and 500 EUR. GPUs with more than 20 GB of memory cost about ten fold. Despite drawbacks in processing speed, it also might make sense to use integrated GPUs with access to huge amounts of DDR4-memory.

5.4 The Workflow

5.4.1 Macro Translation

The CLIJ Fiji plugin and its individual CLIJ operations were programmed in a way which ensures that ImageJ users will easily recognise well-known concepts when translating workflows, and can use CLIJ operations as if they were ImageJ operations. There are some differences, aimed at improved user experience, that we would like to highlight in this section.

The Macro Recorder

The ImageJ macro recorder is one of the most powerful tools in ImageJ. While the user calls specific menus to process images, it automatically records code. The recorder is launched from the menu *Plugins -> Macros -> Record...*. The user can also call any CLIJ operation from the menu. For example, the first step in the nucleus segmentation workflow is to apply a Gaussian blur to a 2D image. This operation can be found in the menu *Plugins > ImageJ on GPU (CLIJ2) > Filter > Gaussian blur 2D on GPU*. When executing this command, the macro recorder will record this code:

```
run("CLIJ2 Macro Extensions", "cl_device=[Intel(R) UHD Graphics
↪    620]");

// gaussian blur
image1 = "NPCsingleNucleus.tif";
Ext.CLIJ2_push(image1);
image2 = "gaussian_blur-1901920444";
sigma_x = 2.0;
sigma_y = 2.0;
```

```
Ext.CLIJ2_gaussianBlur2D(image1, image2, sigma_x, sigma_y);
Ext.CLIJ2_pull(image2);
```

All recorded CLIJ-commands follow the same scheme: The first line initializes the GPU, and explicitly specifies the used OpenCL device while executing an operation. The workflow developer can remove this explicit specification as introduced in ► Sect. 5.3.1. Afterwards, the parameters of the command are listed and specified. Input images, such as *image1* in the example above, are pushed to the GPU to have them available in its memory. Names are assigned to output image variables, such as *image2*. These names are automatically generated and supplemented with a unique number in the name. The developer is welcome to edit these names to improve code readability. Afterwards, the operation *GaussianBlur2D* is executed on the GPU. Finally, the resulting image is pulled back from GPU memory to be visualized on the screen as an image window.

Fiji's Search Bar

As ImageJ and CLIJ come with many commands and huge menu structures, a user may not know in which menu specific commands are listed. To search for commands in Fiji, the Fiji search bar is a convenient tool; it is shown in ◘ Fig. 5.5a. For example, the next step in our workflow is to segment the blurred image using a histogram-based (Otsu's) thresholding algorithm, (Otsu, 1979). When entering *Otsu* in the search field, related commands will be listed in the search result. Hitting the *Enter* key or clicking the *Run* button will execute the command as if it was called from the menu. Hence, also identical code will be recorded in the macro recorder.

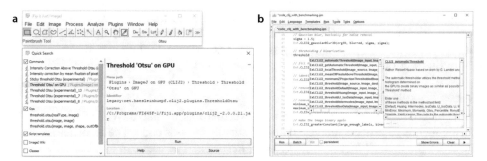

◘ **Fig. 5.5** (a) While recording macros, the Fiji search bar helps to find CLIJ commands in the menu. (b) Auto-Completion in Fiji's script editor supports a workflow developer in finding suitable commands and offers their documentation

The Script Editor and the Auto-Complete Function

In the Macro Recorder window, there is a *Create*-button which opens the Script Editor. In general, it is recommended to record a rough workflow. To extend code, to configure parameters, and to refine execution order, one should switch to the Script Editor. The script editor exposes a third way for exploring available commands: The auto-complete function, shown in ◘ Fig. 5.5b. Typing *threshold* will open two windows: A list of commands which contain the searched word. The position of the searched word within the command does not matter. Thus, entering *threshold* or *otsu* will both lead to the command *thresholdOtsu*. Furthermore, a second window will show the documentation of the respectively selected command. By hitting the *Enter* key, the selected command is auto-completed in the code, for example like this:

```
Ext.CLIJ2_thresholdOtsu(Image_input, Image_destination);
```

The developer can then replace the written parameters *Image_input* and *Image_destination* with custom variables.

The CLIJ website and API Reference

Furthermore, the documentation window of the auto-complete function is connected to the API reference section of the CLIJ website,[3] as shown in ◘ Fig. 5.6. The website provides a knowledge base, holding a complete list of operations and typical workflows connecting operations with each other. For example, this becomes crucial when searching for the CLIJ analog of the ImageJ's Particle Analyzer, as there is no such operation in CLIJ. The website lists typical operations following Otsu thresholding, for example connected component labelling, the core algorithm behind ImageJ's Particle Analyzer.

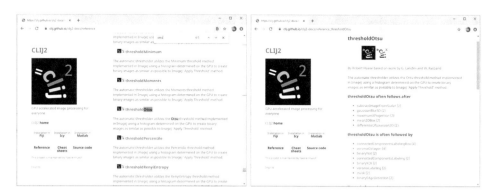

◘ **Fig. 5.6** The online API reference can be explored using the search function of the internet browser, e.g. for algorithms containing *Otsu* (left). The documentation of specific commands contains a list of typical predecessors and successors (right). For example, thresholding is typically followed by connected component labelling, the core algorithm behind ImageJ's Particle Analyzer

---

3 ▶ https://clij.github.io/clij2-docs/reference.

> **❓ Exercise 1**
>
> Open the Macro Recorder and the example image *NPCsingleNucleus.tif*. Type *Otsu* into the Fiji search bar. Select the CLIJ2 method on GPU and run the thresholding using the button *Run*. Read in the online documentation which commands are typically applied before Otsu thresholding. Which of those commands can be used to improve the segmentation result?

5.4.2 The New Workflow Routine

While reconstructing the workflow, this tutorial follows the routines of the classic macro, and restructures the execution order of commands to prevent minor issues with pre-processing before thresholding. The processed dataset is a four-dimensional dataset, consisting of two spatial dimensions, X and Y, channels and frames. When segmenting the nuclear envelope in the original workflow, the first operation applied to the dataset is a *Gaussian blur*:

```
run("Gaussian Blur...", "sigma=1.50 stack");
```

The *stack* parameter suggests that this operation is applied to all time points in both channels, potentially harming later intensity measurements. However, for segmentation of the nuclear envelope in a single time point image, this is not necessary. As discussed in ▶ Sect. 5.3.2, data of this type is not of granular nature and have to be decomposed into 2D images before applying CLIJ operations. We can use the method *pushCurrentSlice* to push a single 2D image to the GPU memory. Then, a 2D segmentation can be generated, utilizing a workflow similar to the originally proposed workflow. Finally, we pull the segmentation back as ROI and perform statistical measurements using classic ImageJ. Thus, the content of the *for*-loop in the original program needs to be reorganized:

```
for (i = 0; i < frames; i ++) {

    // navigate to a given time point in our stack
    Stack.setFrame(i + 1);

    // select the channel showing nuclei
    Stack.setChannel(nuclei_channel);

    // get a single-channel slice
    Ext.CLIJ2_pushCurrentSlice(orgName);

    // segment the nuclear envelope
    nucrimID = nucseg( orgName );

    // select the channel showing nuclear envelope signal
    Stack.setChannel(channel_to_measure);

    // pull segmented binary image as ROI from GPU
    Ext.CLIJ2_pullAsROI(nucrimID);

    // analyze it
    run("Measure");
```

```
      // remove selection
      run("Select None");
}
```

The function *nucseg* takes an image from the nucleus channel and segments its nuclear envelope. ◻ Table 5.1 shows translations from original ImageJ macro functions to CLIJ operations.

While the translation of commands for thresholding is straightforward, other translations need to be explained in more detail, for example the *Analyze Particles* command:

```
run("Analyze Particles...", "size=800-Infinity pixel
↪   circularity=0.00-1.00 show=Masks display exclude clear include
↪   stack");
```

The advanced ImageJ macro programmer knows that this line does post-processing of the thresholded binary image, and executes in fact five operations: (1) It identifies individual objects in the binary image—the operation is known as connected component labeling; (2) It removes objects smaller than 800 pixels (*size=800-Infinity pixel*); (3) It removes objects touching the image edges (*exclude*); (4) It fills black holes in white areas (*include*); and finally (5) it again converts the image to a binary image (*show=Masks*). The remaining parameters of the command, *circularity=0.00−1.00*, *display*, and *clear*, are not relevant for this processing step, or in case of *stack*, specify that the operations should be applied to the whole stack slice-by-slice. Thus, the parameters specify commands which should be executed, but they are not given in execution order. As explained in ▶ Sect. 5.3.2, CLIJ operations are granular: When working with CLIJ, each of the five operations listed above must be executed, and in the right order. This leads to longer code, but also the code which is easier to read and to maintain:

```
// Fill black holes in white objects
Ext.CLIJ2_binaryFillHoles(thresholded, holes_filled);

// Identify individual objects
Ext.CLIJ2_connectedComponentsLabelingBox(holes_filled, labels);

// Remove objects which touch the image edge
Ext.CLIJ2_excludeLabelsOnEdges(labels, labels_wo_edges);

// Exclude objects smaller than 800 pixels
minimum_size = 800;
maximum_size = 1000000; // large number
Ext.CLIJx_excludeLabelsOutsideSizeRange(labels_wo_edges,
↪   large_labels, minimum_size, maximum_size);

// generate a new binary image
Ext.CLIJ2_greaterConstant(large_labels, binary_mask, 0);
```

Finally, the whole translated workflow becomes.[4]

```
1  // configure channels
2  nuclei_channel = 1;
```

---

4 ▶ https://github.com/NEUBIAS/neubias-springer-book-2021/tree/master/Ch05_GPU-accelerating_ImageJ_Macro_image_processing_workflows_using_CLIJ/code/code_clij_final.ijm.

Table 5.1 Translations of ImageJ macro to CLIJ macro, in the context of the example workflow

ImageJ Macro	ImageJ + CLIJ Macro
Gaussian Blur	
`run("Gaussian Blur...", "sigma=1.50 stack");`	`sigma = 1.5;` `Ext.CLIJ2_gaussianBlur2D(orgID, blurred, sigma, sigma);`

Thresholding (Otsu, 1979) and analyze particles to eliminate small objects

ImageJ Macro	ImageJ + CLIJ Macro

`setAutoThreshold("Otsu dark");`
`setOption("BlackBackground", true);`
`run("Convert to Mask", "method=Otsu`
`background=Dark calculate black");`
	`Ext.CLIJ2_thresholdOtsu(blurred, thresholded);`

`run("Analyze Particles...", "size=800-`
`Infinity pixel circularity=0.00-1.00`
`show=Masks display exclude clear`
`include stack");`
	`Ext.CLIJ2_binaryFillHoles(thresholded, holes_filled);`
	`Ext.CLIJ2_connectedComponentsLabelingBox(holes_filled, labels);`
	`Ext.CLIJ2_excludeLabelsOnEdges(labelled, labels_wo_edges);`
	`Ext.CLIJx_excludeLabelsOutsideSizeRange(labels_wo_edges,`
	`large_enough_labels, 800, 1000000);`
	`Ext.CLIJ2_greaterConstant(large_enough_labels, binary_mask, 0);`

Dilation	

`run("Options...", "iterations=2 count=1`
`black edm=Overwrite do=Nothing");`
`run("Dilate", "stack");`
	`radius = 2;`
	`Ext.CLIJ2_maximum2DBox(binary_mask, dilateID, radius, radius);`

Erosion	
`run("Erode", "stack");`	`Ext.CLIJ2_minimum2DBox(binary_mask, erodeID, radius, radius);`
Image subtraction	
`imageCalculator("Difference create stack",` `dilateID, erodeID);`	`Ext.CLIJ2_subtractImages(dilateID, erodeID, resultID);`
ROI generation	

`selectImage( nucrimID );`
`run("Create Selection");`
`selectImage( c2id );`
`run("Restore Selection");`
	`Ext.CLIJ2_pullAsROI(nucrimID);`

```
3    protein_channel = 2;
4
5    // Initialize GPU
6    run("CLIJ2 Macro Extensions", "cl_device=");
7    Ext.CLIJ2_clear();
8
9    // determine current image
10   orgName = getTitle();
11
12   // configure measurements (on CPU)
13   opt = "area mean centroid perimeter shape integrated display
     ↪    redirect=None decimal=3";
14   run("Set Measurements...", opt);
15
16   getDimensions(width, height, channels, slices, frames);
17   for (i = 0; i < frames; i ++) {
18       // select channel and frame to analyze
19       Stack.setChannel(nuclei_channel);
20       Stack.setFrame(i + 1);
21
22       // get a single-channel slice
23       Ext.CLIJ2_pushCurrentSlice(orgName);
24
25       // segment the nuclear envelope
26       nucrimID = nucseg( orgName );
27
28       // select the channel showing nuclear envelope signal
29       Stack.setChannel(protein_channel);
30
31       // pull segmented binary image as ROI from GPU
32       Ext.CLIJ2_pullAsROI(nucrimID);
33
34       // analyse it
35       run("Measure");
36
37       // reset selection
38       run("Select None");
39   }
40
41   // This function segments the nuclear envelope in the nuclei-channel
42   function nucseg( orgID ){
43       // Gaussian blur, basically for noise removal
44       sigma = 1.5;
45       Ext.CLIJ2_gaussianBlur2D(orgID, blurred, sigma, sigma);
46
47       // thresholding / binarization
48       Ext.CLIJ2_thresholdOtsu(blurred, thresholded);
49
50       // fill holes
51       Ext.CLIJ2_binaryFillHoles(thresholded, holes_filled);
52
53       // identify individual objects
54       Ext.CLIJ2_connectedComponentsLabelingBox(holes_filled, labels);
55
56       // remove objects which touch image border
57       Ext.CLIJ2_excludeLabelsOnEdges(labels, labels_wo_edges);
58
```

```
59    // remove objects out of a given size range
60    minimum_size = 800;
61    maximum_size = 1000000;
62    Ext.CLIJx_excludeLabelsOutsideSizeRange(labels_wo_edges,
      ↪  large_labels, minimum_size, maximum_size);
63
64    // make the image binary again
65    Ext.CLIJ2_greaterConstant(large_labels, binary_mask, 0);
66
67    // dilate
68    radius = 2;
69    Ext.CLIJ2_maximum2DBox(binary_mask, dilateID, radius, radius);
70
71    // erode
72    Ext.CLIJ2_minimum2DBox(binary_mask, erodeID, radius, radius);
73
74    // subtract eroded from dilated image to get a band corresponding
      ↪  to nuclear envelope
75    Ext.CLIJ2_subtractImages(dilateID, erodeID, resultID);
76
77    // return result
78    return resultID;
79  }
```

Further Optimization

So far, we translated a pre-existing segmentation workflow without changing processing steps, and with the goal of replicating results. If processing speed plays an important role, it is possible to further optimize the workflow, accepting that results may be slightly different. Therefore, it is necessary to identify code sections which have a high potential for further optimization. To trace down the time consumption of code sections, we now introduce three more CLIJ commands:

```
Ext.CLIJ2_startTimeTracing();
Ext.CLIJ2_stopTimeTracing();
// here comes the workflow we want to analyze
Ext.CLIJ2_getTimeTracing(time_traces);
print(time_traces);
```

By including these lines at the beginning and the end of a macro, we can trace elapsed time during command executions in the log window, as shown in ◱ Fig. 5.7. In that way, one can identify parts of the code where most of the time is spent. In the case of the implemented workflow, connected component labelling appeared as a bottleneck.

In order to exclude objects smaller than 800 pixels from the segmented image, we need to apply (call) connected component labelling. By skipping this step and accepting a lower quality of segmentation, we could have a faster processing. This leads to a shorter workflow:

```
function nucseg( orgID ){
    // blur the image to get a smooth outline
    sigma = 1.5;
    Ext.CLIJ2_gaussianBlur2D(orgID, blurred, sigma, sigma);

    // threshold it
    Ext.CLIJ2_thresholdOtsu(blurred, thresholded);
```

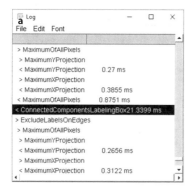
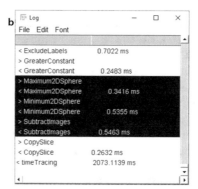

◻ Fig. 5.7 An example of printed time traces reveals that (**a**) connected component labeling takes about 21 ms per slice, whereas (**b**) binary erosion, dilation, and subtraction of images takes about 1.3 ms per slice

```
// fill holes in the binary image
Ext.CLIJ2_binaryFillHoles(thresholded, binary_mask);

// dilate the binary image
radius = 2;
Ext.CLIJ2_maximum2DBox(binary_mask, dilateID, radius, radius);

// erode the binary image
Ext.CLIJ2_minimum2DBox(binary_mask, erodeID, radius, radius);

// subtract the eroded from the dilated image
Ext.CLIJ2_subtractImages(dilateID, erodeID, resultID);
return resultID;
}
```

Analogously, an optimization can also be considered for the classic workflow. When executing the optimized version of the two workflows, we retrieve different measurements, which will be discussed in the following section.

❓ Exercise 2

Start the ImageJ Macro Recorder, open an ImageJ example image by clicking the menu *File > Open Samples > T1 Head (2.4M, 16 bit)* and apply the *Top Hat* filter to it. In the recorded ImageJ macro, activate time tracing before calling the *Top Hat* filter to study what is actually executed when running the *Top Hat* operation and how long it takes. What does the *Top Hat* operation do?

5.4.3 Good Scientific Practice in Method Comparison Studies

When refactoring scientific image analysis workflows, good scientific practice includes quality assurance to check if a new version of a workflow produces identical results, within a given tolerance. In software engineering, the procedure is known as regression testing. Translating workflows for the use of GPUs instead of CPUs, is one such example. In a wider context, other examples are switching major software versions, operating systems, CPU or GPU hardware, or computational environments, such as ImageJ and Python.

Starting from a given dataset, we can execute a reference script to generate reference results. Running a refactored script, or executing a script under different conditions will deliver new results. To compare these results to the reference, we use different strategies, ordered from the simplest to the most elaborated approach: (1) comparison of mean values and standard deviation; (2) correlation analysis; (3) equivalence testing; and (4) Bland-Altman analysis. For demonstration purpose, we will apply these strategies to our four workflows:

— W-IJ: Original ImageJ workflow;
— W-CLIJ: Translated CLIJ workflow;
— W-OPT-IJ: Optimized ImageJ workflow;
— W-OPT-CLIJ: Optimized CLIJ workflow.

In addition, we will execute the CLIJ macros on four computers with different CPU/GPU specifications:

— Intel i5-8265U CPU/ Intel UHD 620 integrated GPU;
— Intel i7-8750H CPU/ NVidia Geforce 2080 Ti RTX external GPU;
— AMD Ryzen 4700U CPU/ AMD Vega 7 integrated GPU;
— Intel i7-7920HQ CPU/ AMD Radeon Pro 560 dedicated GPU;

Comparison of Mean Values and Standard Deviation

An initial and straightforward strategy is to compare mean and standard deviation of the measurements produced by the different workflows. If the difference between then mean measurements exceeds a given tolerance, the new workflow cannot be utilized to study the phenomenon as done by the original workflow. However, if means are equal or very similar, this does not allow us to conclude that the methods are interchangeable. Similar mean and standard deviation values are necessary, but not sufficient to prove method similarity. Results of the method comparison, using mean and standard deviation, are shown in ◻ Table 5.2.

◻ **Table 5.2** Mean ± standard deviation of measured signal intensities resulting from the different considered workflows and different CPU/GPU specifications

Workflow	Intel CPU Intel iGPU	Intel CPU NVidia eGPU	AMD CPU AMD iGPU	Intel CPU AMD dGPU
W-IJ	47.72 ± 3.85	47.72 ± 3.85	47.72 ± 3.85	47.72 ± 3.85
W-CLIJ	47.39 ± 3.64	47.39 ± 3.64	47.74 ± 3.89	47.74 ± 3.89
W-OPT-IJ	46.19 ± 3.9	46.19 ± 3.9	46.19 ± 3.9	46.19 ± 3.9
W-OPT-CLIJ	46.64 ± 3.62	46.64 ± 3.62	47.01 ± 3.87	47.01 ± 3.87

Correlation Analysis

If two methods are supposed to measure the same parameter, they should produce quantitative measurements with high correlation on the same data set. To quantify the level of correlation, Pearson's correlation coefficient r can be utilized. When evaluated on our data, r values were in all cases above 0.98, indicating high correlation. These results are typically visualised by scatter plots, as shown in ◘ Fig. 5.8. Again, high correlation is necessary, but not sufficient, for proving method similarity.

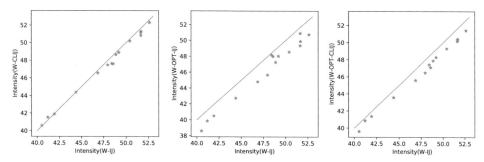

◘ **Fig. 5.8**   Scatter plots of measurements resulting from the original ImageJ macro workflow versus the CLIJ workflow (left), the optimized ImageJ workflow (center), and the optimized CLIJ workflows (right). The orange line represents identity

Equivalence Testing

For proving that two methods A and B result in equal measurements with given tolerance, statistical hypothesis testing should be used. A paired t-test indicates if the observed differences are significant. Thus, a failed t-test is also necessary, but not sufficient to prove method similarity. A valid method for investigating method similarity is a combination of two one-sided paired t-tests (TOST). First, we define a lower and an upper limit of tolerable differences between method A and B, for example ±5%. Then, we apply the first one-sided paired t-test to check if measurements of method B are less than 95% compared to method A, and then the second one-sided t-test to check if measurements of method B are greater than 105% compared to method A. Comparing the original workflow (W-IJ) to the translated CLIJ workflow (W-CLIJ), the TOST showed that observed differences are within the tolerance (p-value < 1e-11).

Bland-Altman Analysis

Another method of analysing differences between two methods is to determine a confidence interval, as suggested by Altman and Bland (1983). Furthermore, so-called Bland-Altman plots deliver a visual representation of differences between methods, as shown in ◘ Fig. 5.9. When comparing the original workflow (W-IJ) to the CLIJ version (W-CLIJ), the mean difference appears to be close to 0.4, and the differences between the methods are within the 95% confidence interval [−0.4, 1]. The means of the two methods range between 40 and 53. Thus, when processing our example

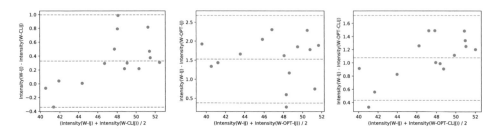

Fig. 5.9 Bland-Altman plots of differences between measurements, resulting from the original ImageJ macro workflow (W-IJ) versus (left) the CLIJ workflow (W-CLIJ), (center) the optimized ImageJ workflow (W-OPT-IJ), and (right) the optimized CLIJ workflows (W-OPT-CLIJ). The dotted lines denote the mean difference (center) and the upper and lower bound of the 95% confidence interval

dataset, the CLIJ workflow (W-CLIJ) delivered intensity measurements of about 1% lower than the original workflow (W-IJ).

5.4.4 Benchmarking

After translating the workflow and assuring that the macro executes the right operations on our data, benchmarking is a common process to analyze the performance of algorithms.

Fair Performance Comparison

When investigating GPU-acceleration of image analysis procedures, it becomes crucial to obtain a realistic picture of the workflows performance. By measuring the processing time of individual operations on GPUs compared to ImageJ operations using CPUs, it was shown that GPUs typically perform faster than CPUs (Haase et al., 2020). However, pushing image data to the GPU memory and pulling results back take time. Thus, the transfer time needs to be included when benchmarking a workflow. The simplest way is to measure the time at the beginning of the workflow and at its end. Furthermore, it is recommended to exclude the needed time to load from hard drives, assuming that this operation does not influence the processing time of CPUs or GPUs. After the *open( )* image statement, the initial time measurement should be inserted:

```
start_time = getTime();
```

Before saving the results to disc, we measure the time again and calculate the time difference:

```
end_time = getTime();
print("Processing took " + (end_time-start_time) + " ms");
```

The *getTime( )* method in ImageJ delivers the number of milliseconds since midnight of January 1, 1970 UTC. By subtracting two subsequent time measurements, we can calculate the passed time in milliseconds.

Warm-up Effects

To ensure reliable results, time measurements should be repeated several times. As shown in ▶ Sect. 5.3, the first execution of a workflow is often slower than subsequent runs. The reason is the so-called warm-up effect, related to just-in-time (JIT) compilation of Java and OpenCL code. This compilation takes time. To show the variability of measured processing times between the original workflow and the CLIJ translation, we executed all the considered workflows in loops for 100 times each. To eliminate resulting effects of different and subsequently executed workflows, we restarted Fiji after each 100 executions. From the resulting time measurements, we derived a statistical summary in a form of the median speedup factor. Visualized by box plots, we have generated an overview of the performance of the four different workflows, executed on four tested systems.[5]

Benchmarking Results and Discussion

The resulting overview of the processing times is given in ◻ Fig. 5.10. Depending on the tested system, the CLIJ workflow results in median speedup factors between 1.5 and 2.7. These results must be interpreted with care. As shown in (Haase et al., 2020), workflow performance depends on many factors, such as the number of operations and parameters, used hardware, and image size. When working on small images, which fit into the so-called Levels 1 and 2 cache of internal CPU memory, CPUs typically outperform GPUs. Some operations perform faster on GPUs, such as convolution, or other filters which take neighboring pixels into account. By nature, there are operations which are hard to compute on GPUs. Such an example is the connected component labelling. As already described in ▶ Sect. 5.4.2, we identified this operation as a bottleneck in our here considered example workflow. Without this operation, the optimized CLIJ workflow performed up to 5.5 times faster than the original. Hence, a careful workflow design is a key to high performance. Identifying slow parts of the workflow and replacing them with alternative operations becomes routine when processing time is a relevant factor.

❓ Exercise 3

Use the methods introduced in this section to benchmark the script presented in ▶ Sect. 5.3. Compare the performance of the *mean* filter in ImageJ with its CLIJ counterpart. Determine the median processing time of both filters, including push and pull commands when using CLIJ.

5.5 **Summary**

The method of live-cell imaging, in particular recording long-term time-lapses with high spatial resolution, is of increasing importance to study dynamic biological processes. Due to increased processing time of such data, image processing may become the major bottleneck. In this chapter, we introduced one potential solution for faster

---

5 ▶ https://github.com/NEUBIAS/neubias-springer-book-2021/blob/master/Ch05_GPU-accelerating_ImageJ_Macro_image_processing_workflows_using_CLIJ/code/performance_comparison.ipynb.

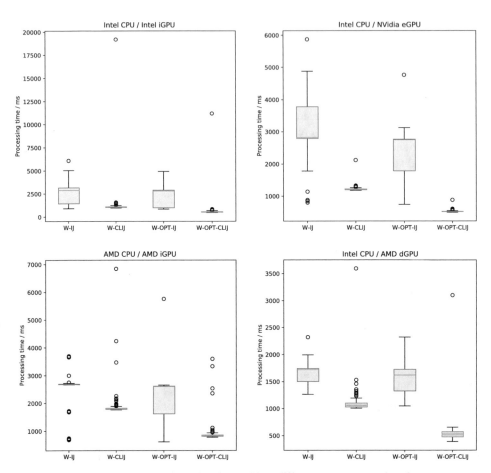

Fig. 5.10 Box plots showing processing times of four different macros, tested on four computers. In the case of the of classic ImageJ macro, blue boxes range from the 25th to the 75th percentile of processing time. Analogously, green boxes represent processing times of the CLIJ macro. The orange line denotes the median processing time. Circles denote outliers. In case of the CLIJ workflow, outliers typically occur during the first iteration, where compilation time causes the warm-up effect

processing, namely by GPU-accelerated image processing using CLIJ. We also demonstrated a step-by-step translation of a classic ImageJ Macro workflow to GPU-accelerated macro workflow. Clearly, GPU-acceleration is suited for particular use cases. Typical cases are

- processing of data larger than 10 MB per time point and channel;
- application of 3D image processing filters, such as convolution, mean, minimum, maximum, Gaussian blur;
- need for acceleration of workflows which take significant amount of time, especially if processing is 10 times longer than loading and saving images;
- extensive workflows with multiple operations, consecutively executed on the GPU;
- last but not least, utilizing sophisticated GPU-hardware with a high memory bandwidth, typically using GDDR6 memory.

When these needs/conditions are met, speedup factors of one or two orders of magnitude are feasible. Furthermore, the warm-up effect is crucial. For example, if the first execution of a workflow takes ten times longer than subsequent executions, it becomes obvious that at least 11 images have to be processed to overcome the effect and to actually save time. When translating a classic workflow to CLIJ, some refactoring is necessary to follow the concept of processing granular units of image data by granular operations. This also improves readability of workflows, because operations on images are stated explicitly and in the order of execution. Additionally, the shown methods for benchmarking and quality assurance can also be used in different scenarios, as they are general method comparison strategies. GPU-accelerated image processing opens the door for more sophisticated image analysis in real-time. If days of processing time can be saved, it is worth investing hours required to learn CLIJ.

Solutions to the Exercises

✅ Exercise 1

While applying image processing methods, the ImageJ Macro recorder records corresponding commands. This offers an intuitive way to learn ImageJ Macro programming and CLIJ. After executing this exercise, the recorder should contain code like this:

```
open("/path/to/images/NPCsingleNucleus.tif");
selectWindow("NPCsingleNucleus.tif");
run("CLIJ2 Macro Extensions", "cl_device=[Intel(R) HD Graphics
↪    630]");

// threshold otsu
image1 = "NPCsingleNucleus.tif";
Ext.CLIJ2_push(image1);
image2 = "threshold_otsu-936068520";
Ext.CLIJ2_thresholdOtsu(image1, image2);
Ext.CLIJ2_pull(image2);
```

It opens the dataset, initializes the GPU, pushes the image to GPU memory, thresholds the image, and pulls the resulting image back to show it on the screen.

The Fiji search bar allows to select CLIJ methods. The corresponding dialog gives access to the CLIJ website, where the user can read about typical predecessor and successor operations. For example, as shown in ▶ Sect. 5.4.1 in ◻ Fig. 5.6, operations such as Gaussian blur, Mean filter, and Difference-Of-Gaussian are listed, which allow an improved segmentation, because they reduce noise.

✅ Exercise 2

The recorded macro, adapted to print time traces, looks like this:

```
run("T1 Head (2.4M, 16-bits)");
run("CLIJ2 Macro Extensions", "cl_device=[Intel(R) UHD Graphics
↪    620]");

// top hat
image1 = "t1-head.tif";
Ext.CLIJ2_push(image1);
image2 = "top_hat-427502308";
radius_x = 10.0;
```

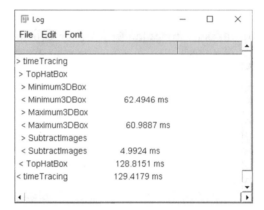

◻ Fig. 5.11   While executing the *Top Hat* filter, activated time tracing reveals that this operation consists of three subsequently applied operations: a *minimum* filter, a *maximum* filter and image subtraction

```
radius_y = 10.0;
radius_z = 10.0;

// study time tracing of the Top Hat filter
Ext.CLIJ2_startTimeTracing();
Ext.CLIJ2_topHatBox(image1, image2, radius_x, radius_y, radius_z);
Ext.CLIJ2_stopTimeTracing();

Ext.CLIJ2_pull(image2);

// determine and print time traces
Ext.CLIJ2_getTimeTracing(time_traces);
print(time_traces);
```

The traced times, while executing the *Top Hat* filter on the T1-Head dataset, are shown in ◻ Fig. 5.11. The *Top Hat* filter is a *minimum* filter applied to the original image, which is followed by a *maximum* filter. The result of these two operations is subtracted from the original. The two filters take about 60 ms each on the 16 MB large input image, the subtraction takes 5 ms. The *Top Hat* filter altogether takes 129 ms. *Top hat* is a technique to subtract background intensity from an image.

✓ Exercise 3

For benchmarking the mean 3D filter in ImageJ and CLIJ two example macros are provided online.[6] We executed them on our test computers and determined median execution times between 1445 and 5485 ms for the ImageJ filter and from 81 to 159 ms for the CLIJ filter, respectively.

---

6   ► https://github.com/NEUBIAS/neubias-springer-book-2021/tree/master/Ch05_GPU-accelerating_ImageJ_Macro_image_processing_workflows_using_CLIJ/code/exercise_3.

Take-Home Message

In this chapter you learned how a classic ImageJ macro can be translated to a GPU-accelerated CLIJ macro. Image processing on a CPU might become time-consuming, especially when processing large datasets, such as complex time-lapse data. Therefore, it is important to rethink parts of the workflow and to speed it up by forwarding processing tasks to a GPU. For an optimal exploitation of the computing power of GPUs, it is recommended to process data time-point by time-point, and also not to apply filters to the whole time-lapse at once. Furthermore, we introduced strategies for a good scientific practice on benchmarking and quantitative comparison of results between an original and a GPU-accelerated workflow to assure that the GPU-accelerated workflow performs with equal measurement results and under a given tolerance.

Acknowledgements We would like to thank Kota Miura and Andrea Boni for sharing their image data and code openly with the community. It was the base for our chapter. We also thank Dominic Waithe (University of Oxford), Tanner Fadero (UNC Chapel Hill), Anna Hamacher (Heinrich-Heine-Universität Düsseldorf), Johannes Girstmair (MPI-CBG) and Thomas Brown (CSBD/MPI-CBG) for proofreading and providing feedback. We thank Gene Myers (CSBD/MPI-CBG) for constant support and giving us the academic freedom to advance GPU-accelerated image processing in Fiji. We also would like to thank our colleagues who supported us in making CLIJ and CLIJ2 possible in first place, namely Alexandr Dibrov (CSBD/MPI-CBG), Brian Northon (True North Intelligent Algorithms) Deborah Schmidt (CSBD/MPI-CBG), Florian Jug (CSBD/MPI-CBG, HT Milano), Loïc A. Royer (CZ Biohub), Matthias Arzt (CSBD/MPI-CBG), Martin Weigert (EPFL Lausanne), Nicola Maghelli (MPI-CBG), Pavel Tomancak (MPI-CBG), Peter Steinbach (HZDR Dresden), and Uwe Schmidt (CSBD/MPI-CBG). Furthermore, development of CLIJ is a community effort. We would like to thank the NEUBIAS Academy (▶ https://neubiasacademy. org/.) and the Image Science community (▶ https://image.sc/.) for constant support and feedback. R.H. was supported by the German Federal Ministry of Research and Education (BMBF) under the code 031L0044 (Sysbio II) and by the Deutsche Forschungsgemeinschaft (DFG, German Research Foundation) under Germany's Excellence Strategy—EXC2068—Cluster of Excellence Physics of Life of TU Dresden.

Further Readings On top of the given references in the main text, readers interested in state-of-the-art benchmarking approaches in high performance computing are recommended to read the overview given by Hoefler and Belli (2015). Furthermore, a research software engineers perspective on developing GPU-accelerated applications is also worth taking a closer look (van Werkhoven et al., 2020).

References

Altman DG, Bland JM (1983) Measurement in medicine: The analysis of method comparison studies. J R Stat Soc Ser D 32(3):307–317. https://doi.org/10.2307/2987937. https://rss.onlinelibrary.wiley.com/doi/abs/10.2307/2987937

Boni A, Politi AZ, Strnad P, Xiang W, Hossain MJ, Ellenberg J (2015) Live imaging and modeling of inner nuclear membrane targeting reveals its molecular requirements in mammalian cells. J Cell Biol 209(5):705–720. https://doi.org/10.1083/jcb.201409133. https://rupress.org/jcb/article-pdf/209/5/705951675/jcb_201409133.pdf

Haase R, Royer LA, Steinbach P, Schmidt D, Dibrov A, Schmidt U, Weigert M, Maghelli N, Tomancak P, Jug F, Myers EW (2020) CLIJ: GPU-accelerated image processing for everyone. Nat Methods 17(1):5–6. https://doi.org/10.1038/s41592-019-0650-1

Hoefler T, Belli R (2015) Scientific benchmarking of parallel computing systems: twelve ways to tell the masses when reporting performance results. In: SC '15: Proceedings of the international conference for high performance computing, networking, storage and analysis, p 1–12

Khronos-Group (2020) The open standard for parallel programming of heterogeneous systems. https://www.khronos.org/opencl/. Accessed 12 Aug 2020

Miura K (2020) Measurements of intensity dynamics at the periphery of the nucleus. Springer, Cham, p 9–32. https://doi.org/10.1007/978-3-030-22386-1_2

Otsu N (1979) A threshold selection method from gray-level histograms. IEEE Trans Syst Man Cybern 9(1):62–66

Schindelin J, Arganda-Carreras I, Frise E, Kaynig V, Longair M, Pietzsch T, Preibisch S, Rueden C, Saalfeld S, Schmid B (2012) Fiji: an open-source platform for biological-image analysis. Nat Methods 9(7):676—82. https://doi.org/10.1038/nmeth.2019

Schneider CA, Rasband WS, Eliceiri KW (2012) NIH image to imageJ: 25 years of image analysis. Nat Methods 9(7):671

van Werkhoven B, Palenstijn WJ, Sclocco A (2020) Lessons learned in a decade of research software engineering GPU applications. In: Krzhizhanovskaya VV, Závodszky G, Lees MH, Dongarra JJ, Sloot PMA, Brissos S, Teixeira J (eds) Computational science–ICCS 2020. Springer, Cham, pp 399–412

How to Do the Deconstruction of Bioimage Analysis Workflows: A Case Study with SurfCut

Marion Louveaux and Stéphane Verger

Contents

This Chapter has been reviewed by Mafalda Sousa, I3S - Advanced Light Microscopy, University of Porto.

© The Author(s) 2022
K. Miura, N. Sladoje (eds.), *Bioimage Data Analysis Workflows–Advanced Components and Methods*, Learning Materials in Biosciences, https://doi.org/10.1007/978-3-030-76394-7_6

What You Will Learn in This Chapter

Published bioimage analysis workflows are designed for a specific biology use case and often hidden in the material and methods section of a biology paper. The art of the bioimage analyst is to find these workflows, deconstruct them and tune them to a new use case by replacing or modifying components of the workflow and/or linking them to other workflows.

In this chapter, you will learn how to adapt a published workflow to your needs. More precisely, you will learn how to: deconstruct a bioimage analysis workflow into components; evaluate the fit of each component to your needs; replace one element by another one of your choice; benchmark this new workflow against the original one; and link it to another workflow. Our target for workflow deconstruction is *SurfCut*, an ImageJ macro for the projection of 3D surface tissue.[1]

6.1 Introduction

6.1.1 A Workflow and Its Components

Bioimage analysis workflows and components are defined as follows (Miura and Tosi, 2016): (1) A workflow is a set of components assembled in some specific order to process biological images and estimate some numerical parameters relevant to the biological system under study; (2) Components are implementations of certain image processing and analysis algorithms. Each component alone does not solve a bioimage analysis problem. Components may take forms of a single menu item in image processing software, a plugin, a module, an add-on, or a class in an image processing library. Workflows take image data as input, and output either processed images or numerical values. A workflow can be a combination of components from the same or different software packages and can, for example, come under the form of a script that calls components in a sequence, or a detailed step-by-step instruction on how to chain a sequence of components (Miura and Tosi, 2017; Miura et al., 2020).

6.1.2 What Is Deconstruction?

Bioimage Analysis Workflows are designed for specific purpose, so usually, they cannot be used as a general tool for different problems. Then how can we learn how to create bioimage analysis workflows? One way is to do everything from scratch. Another way is to learn from other bioimage analysis workflows, modify them, and reassemble components to create something new for a specific purpose. We call this (a workflow) "*deconstruction*". The process of deconstruction was initially proposed by Jacques Derrida, a French philosopher, as a criticism against the modern philosophy. Instead of constructing ideas, which implicitly builds on hidden but solid principles as the base of such construction, deconstruction is a way of shifting ideas by crit-

---

1   This chapter was communicated by Mafalda Sousa, I3S—Advanced Light Microscopy, University of Porto, Portugal.

ical thinking, sometimes denial, and in other times the restructuring of preexisting principles.

The deconstruction of bioimage analysis workflow was introduced as a pedagogic method for the Bioimage Analyst School of NEUBIAS. Deconstructing a workflow means identifying and isolating each of its components in order to assess their quality and possibly replace them with more suitable components. In addition to using it as a powerful pedagogical tool, one of the main interest in deconstructing a workflow is to avoid spending time and effort "re-inventing the wheel", and instead to re-use, optimize or adapt an existing method to the new users' needs.

6.1.3 A Case of Study of Workflow Deconstruction: SurfCut

The ImageJ macro "SurfCut" was chosen as a study case for workflow deconstruction during the NEUBIAS training school TS15 (Bordeaux, March 2020). Interestingly, this led to numerous new ideas and ways to implement SurfCut. Some trainees added GPU processing capability with CLIJ (Haase et al., 2020), while others completely re-wrote the workflow in Python[2] and Matlab[3] and benchmarked the different versions (SurfCut, GPU-SurfCut, Python-SurfCut and Matlab-SurfCut). Furthermore, this deconstruction session, along with the writing of this book chapter, also prompted us to develop a new version of the SurfCut macro, SurfCut2, including a complete refactoring of the code (as described in this book chapter), bug-fixing, and addition of new functions.[4] In this chapter, we explain in detail the procedure for workflow deconstruction based on these experiences, using SurfCut as an example target workflow.

6.1.4 What Is SurfCut?

SurfCut is an ImageJ macro that allows the numerical extraction of a thin, curved, layer of signal in a 3D confocal stack by taking as reference the surface of a 3D biological object present in the volume of the stack (Erguvan et al., 2019). The macro is written in the ImageJ1.x(IJ1) macro language, and runs on the Fiji platform (Schindelin et al., 2012). Using built-in ImageJ functions, the biological object in the image is blurred, segmented, filled, shifted in the Z-axis at two different depths and used as a mask to erase unwanted raw signals at a chosen distance from the surface of the detected object (◘ Fig. 6.1, and detailed description in ▸ Sect. 6.4). The whole workflow can be viewed as a sort of "object surface"-guided signal filtering method. This allows the removal of unwanted signals relative to the surface of the biological object and the extraction of specific structures from the 3D stack, such as the cell contours (◘ Fig. 6.1) or outer epidermal cortical microtubules. As such, this workflow has already been incorporated as a component of larger workflows, as a preprocessing step for cell segmentation or cortical microtubule signal quantification (Baral et al., 2021; Erguvan et al., 2019; Takatani et al., 2020).

---

2 ▸ https://pypi.org/project/surfcut/.
3 ▸ https://github.com/martinschatz-cz/surfcut-matlab.
4 ▸ https://github.com/VergerLab/SurfCut2.

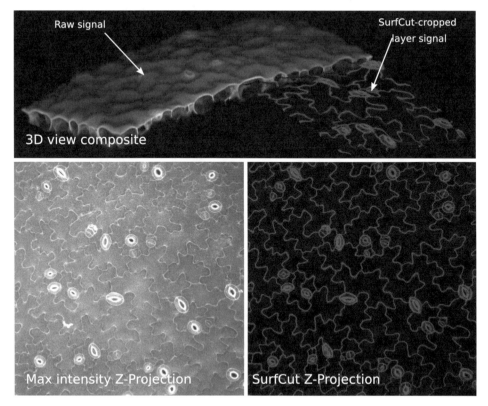

◻ Fig. 6.1 Overview of SurfCut principle and output, applied on *Arabidopsis thaliana* cotyledon epidermal cells stained with propidium iodide and imaged in 3D with a confocal microscope. Top panel is a combination of half of the raw confocal signal (grey) and half of the "SurfCut-extracted" signal (red), partially overlapped and tilted in 3D to show the relationship between the raw signal and output. Bottom left panel is a max-intensity projection of the raw signal. Bottom right panel is a max-intensity projection of the "SurfCut-extracted" signal, highlighting how the process efficiently preserves the cell contour (anticlinal) signal in the epidermal layer while removing signal from the periclinal cell contours

6.1.5 What Was SurfCut Developed for?

SurfCut was originally developed as a pre-processing tool to filter out unwanted signals and perform a Z-projection prior to 2D segmentation of epidermal plant cells. The so-called ''puzzle-shaped pavement cells'' of the leaf epidermis harbor very particular shapes (◻ Fig. 6.1). This is a very interesting system to study the morphogenesis of single cells in a tissue context. To understand how these cell shapes emerge, a proper shape quantification with several genetic backgrounds, or under specific treatment conditions, is required. Many methods were developed to quantify and compare cell shapes based on 2D cell contours (Möller et al., 2017; Sánchez-Corrales et al., 2018; Wu et al., 2016). As the leaf epidermis is a 3D curved surface, a Z projection is required prior to the use of any of these tools. Given the lack of available user-friendly tools to perform a proper extraction of 2D cell contours from 3D confocal stacks, we developed SurfCut (Erguvan et al., 2019).

Although the SurfCut macro was written in the context of a biological project and could have ended (somewhat hidden) in the "Material and method"—section of a larger biological publication (still being finalized at the time of writing this chapter but available as a preprint (Malivert et al., 2021)), we decided to publish it separately (Erguvan et al., 2019), to assign a DOI to the code and provide image data, also identified with a DOI (Erguvan and Verger, 2019), to enable testing of the macro. We think that the publishing of this type of macro gives more visibility to the bioimage analysis workflows and, by giving all the space needed to the description of the workflow, ensures a greater reproducibility.

6.1.6 Other Similar Tools

Before developing SurfCut, we had identified in our bibliographical searches other workflows performing apparently similar outputs, but none of them fitted exactly our needs. As described in Erguvan et al. (2019), we were originally using the software MorphoGraphX (MGX) (Barbier de Reuille et al., 2015) that provides a very accurate solution to our problem (Erguvan et al., 2019; Verger et al., 2018), but requires too many manual steps and does not easily allow batch processing. In addition, Merryproj (Barbier de Reuille et al., 2005), SurfaceProject (Band et al., 2014), LSM-W2 (Zubairova et al., 2019) and Smooth 2D manifold (Shihavuddin et al., 2017) were discussed in Erguvan et al. (2019) and were found inadequate for our purpose. After the independent publication of the SurfCut macro, we discovered other workflows that our first search had not revealed, such as the ImageJ macro identifyuppersurfacev2 (Galea et al., 2018),[5] or the ImageJ plugin MinCostZSurface (Li et al., 2006).[6] We also identified more advanced workflows that would not have fitted our needs for simplicity (Candeo et al., 2016; Heemskerk and Streichan, 2015; Schmid et al., 2013). Furthermore, since the publication of SurfCut, additional workflows, such as the ImageJ plugins Ellipsoid Surface Projection (Viktorinová et al., 2019), SheetMesh-Projection[7] (Wada and Hayashi, 2020) and LocalZProjector (Herbert et al., 2021) were developed to serve a similar purpose. In total, there are at least ten different workflows that can perform the type of signal layer extraction that SurfCut performs. While all these tools allow the generation of relatively similar output, almost all of them use a different approach. In addition, they are tailored to specific needs, such that some of these tools outperform others on a certain type of images, thus offering a large choice of alternative workflow components to perform this specific pre-processing step.

In the following sections, we present how to deconstruct SurfCut (Erguvan et al. 2019), i.e. how to identify its different components in the reference publication and in the code. We then explain how to refactor the code, replace one component and

---

5   ▸ https://www.ucl.ac.uk/child-health/research/core-scientific-facilities-centres/confocal-microscopy/publications see section "Published ImageJ/Fiji macro".

6   ▸ https://imagej.net/Minimum_Cost_Z_surface_Projection.

7   ▸ https://signaling.riken.jp/en/en-tools/imagej/1743/.

benchmark the new workflow against the original one. Finally, we explore how to integrate this workflow with other workflows.

6.2 Dataset

The SurfCut macro was released with test image data of around 535 Mb. This data set was uploaded to Zenodo with a thorough description of the imaging conditions, and identified with its DOI: ▶ http://doi.org/10.5281/zenodo.2577053 (Erguvan and Verger, 2019).

6.3 Tools

- Fiji: Download and install Fiji on your computer   (▶ https://imagej.net/Fiji/Downloads)
- ImageJ macro SurfCut: Download the "SurfCut.ijm" macro file to your computer (▶ https://github.com/sverger/SurfCut). To run the macro in Fiji either click on Plugins>Macro>Run and select "SurfCut.ijm", or drag and drop "SurfCut.ijm" into the Fiji window and click run.
- ImageJ macro SurfCut2: Download the "SurfCut2.ijm" macro file to your computer (▶ https://github.com/VergerLab/SurfCut2). Follow the same instructions as for the ImageJ macro SurfCut.
- ImageJ macro used for exercises in this chapter can be found at: ▶ https://github.com/NEUBIAS/neubias-springer-book-2021

6.4 Workflow

In this section, we propose a step-by-step deconstruction and modification of the SurfCut workflow. The concepts and exercises in each step can be generalised to any kind of bioimage analysis workflow.

We take the following steps for the deconstruction of the workflow:
- Step 1: Identify components in the description of a workflow and in the code;
- Step 2: Draw a workflow scheme;
- Step 3: Identify limitations on input format, processing capabilities, simplicity to re-use;
- Step 4: Identify block of codes corresponding to components;
- Step 5: Refactor code;
- Step 6: Replace a component of the workflow;
- Step 7: Compare the performance of the original workflow with a modified one;
- Step 8: Link this workflow with another workflow.

6.4.1 Step 1. Identification of Components in the Textual Description

When working with a published workflow, the first step is to identify the components in the text of the publication and the order in which they are used. Nowadays, published workflows are often accompanied by a detailed user manual and/or a "readme" if the code is released on GitHub or GitLab. This text can also contain additional information on the components and on the links between them.

? Exercise 1

1. Read Erguvan et al. (2019) and underline in the text all elements describing the components of the SurfCut workflow. Then summarize the result as an ordered list of components.

2. Which additional useful information relative to the components can you find on the GitHub repository of the SurfCut macro[8]?

✓ Solution to Exercise 1

1. All text elements describing the components of the SurfCut macro in Erguvan et al. (2019) are on page 3 in the Methods section, in the "2D cell contour extraction with SurfCut" paragraph:
    - "The macro has two modes: (1) ''Calibrate,'' [...], and (2) ''Batch,'' [...]".
    - "The stack is first converted to 8 bit."
    - "De-noising of the raw signal is then performed using the `Gaussian Blur` function."
    - "The signal is then binarized using the `Threshold` function."
    - "an equivalent of the ''edge detect'' process from MGX[9] is performed [...]; each slice from the binarized stack, starting from the top slice, is successively projected (Z-project) [...]. This ultimately creates a new binary stack in which all the binary signals detected in the upper slices appear projected down on the lower slices, effectively filling the holes in the binary object."
    - "This new stack is then used as a mask shifted in the Z direction, to subtract the signal from the original stack above and below the chosen values depending on the desired depth of signal extraction."
    - "The cropped[10] stack is finally projected along the Z-axis using the maximal fluorescence intensity in order to obtain a 2D image."

    The SurfCut workflow has 6 components: (1) bit-depth conversion, (2) denoising, (3) thresholding and binarization, (4) edge detection, (5) masking, and (6) Z-projection (◻ Fig. 6.2). The workflow can be run one component at a time, to allow for selection of parameters per component (calibrate mode), or automatically (batch mode).

---

8   ▶ https://github.com/sverger/SurfCut.
9   MorphoGraphX.
10   The exact term is "masked".

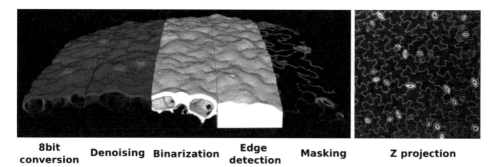

| 8bit conversion | Denoising | Binarization | Edge detection | Masking | Z projection |

▣ **Fig. 6.2** Output of each processing step of the SurfCut workflow

2. In the GitHub repository of the SurfCut macro, a careful reading of the "readme" and user guide[11] identifies and confirms the components found in the text of the publication. Note that dialog boxes to interact with the user are not considered as components of the workflow.

6.4.2 Step 2. Drawing a Workflow Scheme

We identified above the workflow components from the text. Let us now draw a scheme of the workflow. A workflow scheme summarizes and links all the components of a workflow. This scheme will serve as a guide to get an overview of the workflow, and identify those components in the code that can be refined after Step 4, if needed. ▣ Figure 6.3 is a graphical scheme of a general bioimage analysis workflow.

❓ Exercise 2

Utilizing information found in Exercise 1, draw the scheme of the SurfCut workflow: start drawing one box per component following the guidelines in Step 1. Then identify each component by a short informative name and link components with each other, so that the input of a component is an output of the previous component.

Bioimage analysis workflow

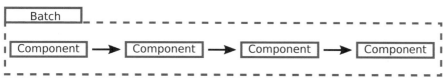

▣ **Fig. 6.3** Workflow scheme example

---

11 ▶ https://github.com/sverger/SurfCut/blob/master/SurfCut_UserGuide.pdf.

SurfCut workflow

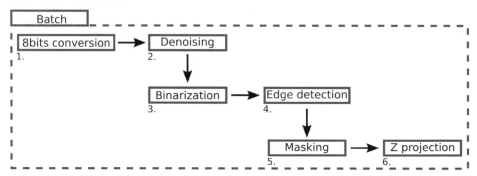

■ **Fig. 6.4**   SurfCut workflow scheme

✔ **Solution to Exercise 2**

The SurfCut macro has two modes: (1) "calibrate", where the components are executed one-by-one and only once, and (2) "batch", where the complete workflow is repeated on several images. We draw a batch component to illustrate the batch mode. We then draw the 6 components of the workflow inside the batch component and link them in the order in which they appear in the text: Component 1 performs the conversion to 8-bit pixel representation, Component 2 performs the denoising, etc. In the text, we will now refer to components using the following wording: Component 1 "8 bits conversion", Component 2 "Denoising", etc (■ Fig. 6.4). For simplicity of the scheme, we ignored import and export components (such as User-Interface for file selection or saving of results). These can be included as well, especially if the import or export components correspond to non-trivial steps (e.g. specific data format).

6.4.3 Step 3. Assessment of Prerequisites and Limitations

In the two previous steps, we identified the components of the workflow described in the publication and drew a workflow scheme. We now have an overview of the workflow and can make more confident assessment on if the workflow is appropriate to solve our biological question or not. To determine if we can use the workflow as it is, if it is sufficient to only change a couple of the components to adapt the workflow to our data, or if the workflow is not adequate at all for our data, we need to make some additional (final) checks:

- Data format compatibility: Is the input format (.tif, .png, .czi...) and type of data (2D, 3D, time-series) that we have compatible with the format and type required by the workflow?
- Processing capacity: Is the amount and size of the data compatible with the workflow (fully manual workflow or very slow workflow versus high content screening data; included calibration step requiring a minimum of 30 images versus 5 images available only...)?

- Data content compatibility: Are the type of biological data and markers that we have to work with compatible with what is considered in the workflow (membrane marker versus nuclear marker, epithelial marker versus whole tissue marker, flat versus curved tissue...)?
- Output adequacy: Will the output data generated by the workflow (new images, numerical values, plots...) be actually useful for what we intend to do (get biological results, benchmark the workflow against another, embed in a larger workflow...)?

If the answer is *no* to one, or several of these questions, the next question to answer is: Could one, or several of the components be replaced by a more adapted or efficient one(s)? Here we assume that the macro language and the use of ImageJ/Fiji is not an obstacle for any bioimage analyst. For other more advanced or less known languages, as well as more exotic software, another sequence of preliminary checks would be:
- Language: In which language is this workflow written?
- Platform: On which platform can I execute it?
- Inter-operability: How complex will it be to link this workflow to my other tools written in another language or executed on another platform?
- Code migration capability: In case I need to make some modifications to the workflow, do I have other options than fully rewriting it in my favorite language?

? Exercise 3

1. Install the SurfCut macro and execute it on the associated data.
2. Identify, based on the text of the reference publication, the "readme" in the Github repository and the user guide, all elements restricting the datasets of certain type to be used with SurfCut.
3. For each use case below, download the dataset and explain if and why the dataset could be processed directly with SurfCut, without any workflow modification, using the checks defined above. We assume that the output of the workflow (a 2D projection) is what we need.

   (a) Use case 1: 3D light sheet microscopy images of a *Tribolium* epithelium (Vorkel et al., 2020).
   Dataset: ▶ https://zenodo.org/record/3981193#.Xzo8pTU6-60,   take "Strausberg_Tribolium_LAGFP_tailpole_runC0opticsprefused301310.tif". This dataset was used as an example to showcase another projection tool.[12]

   (b) Use case 2: 3D confocal and spinning disk microscopy images of *Drosophila* epithelia (notum and wing disc, Valon and Staneva, 2020).
   Dataset: ▶ https://zenodo.org/record/4114074#.X5AJAe06-60. Take the image named "notum2_GFP.tif".

   (c) Use case 3: 3D light sheet microscopy images of single cells (Driscoll et al., 2019).
   Dataset: ▶ https://cloud.biohpc.swmed.edu/index.php/s/Z9j62w2FCareyJY/download), in the folder called "testData". Each image is associated with a text file describing the imaging conditions (AcqInfo.txt). There are three examples of MV3   melanoma   cells   ("krasMV3")   and   one   example of conditionally immortalized hematopoietic precursors to dendritic

---

12 ▶ https://clij.github.io/assistant/sphere_projection.

cells ("lamDendritic"). Associated GitHub repository: ▶ https://github.com/DanuserLab/u-shape3D and research article: ▶ https://www.ncbi.nlm.nih.gov/pmc/articles/PMC7238333/.

(d) Use case 4: light sheet image of a gastric cancer spheroid (Rocha et al., 2020). Dataset: ▶ https://zenodo.org/record/4244952#.X6L_ZIj7SHs.

✅ Solution to Exercise 3

1. See ▶ Sects. 6.2 and 6.3, as well as the installation instructions and the userguide on the GitHub repository.[13]

2. Prerequisites

   Pre-requisites and limitations found in the main text of the publication:

   − "the acquired signal must be strong and continuous enough at the edge of the sample in order for the signal to be detected and segmented from the background noise by a simple conversion to a binary image." (Methods section, in "Confocal microscopy")

   − "avoid the presence of artifacts, e.g., from stained cell debris or bacteria at the surface of the sample." (Methods section, in "Confocal microscopy")

   − "The first slice of the stack should be the top surface of the sample in order for the process to work properly." (Methods section, in "2D cell contour extraction with SurfCut")

   − "a new method (SurfCut) to extract cell contours or specific thin layers of a signal at a distance from the surface of samples in 3D confocal stacks." (Results and discussion section, "2D cell contour extraction from 3D samples with MGX[14] and SurfCut")

   − "the associated error can become important for samples with high curvature." (Results and discussion section, "2D cell contour extraction from 3D samples with MGX and SurfCut")

   − "In principle, this tool may be used on any 3D stack (e.g., confocal or light-sheet microscopy) originating from either animal, fungi, or plant systems." (Conclusions section)

   − "SurfCut is particularly well suited for tissues with a low curvature " (Conclusions section)

   − "SurfCut is very well suited for high-throughput pavement cell contour extraction and further quantification. [...] Besides, SurfCut can also be used to extract other types of signals, such as cortical microtubules, allowing a suppression of the background noise coming from the signal below." (Conclusions section)

   − "SurfCut can be a very useful tool for the 2D representation (from image-based screening protocols to publication figures) of 3D confocal data in which overlapping signal from different depths in the stack hinders the visualization of signal or structures of interest." (Conclusions section)

   Prerequisites and limitations found in the "readme" of the GitHub repository:[15]

---

13 ▶ https://github.com/sverger/SurfCut.
14 MorphoGraphX.
15 ▶ https://github.com/sverger/SurfCut/blob/master/README.md.

- "This can, for example, be used to extract the cell contours of the epidermal layer of cells." (Description section)
- "SurfCut [...] is in principle only adequate for sample with a relatively simple geometry." (How it works section)
- "3D confocal stacks in .tif format, in which the top of the stack should also be the top of the sample." (Prerequisites section)

Prerequisites and limitations found in the user guide:[16]
- "Our image analysis pipeline was developed to extract cell contours or specific layers of signal in confocal images of plant samples, but can in principle be used on any 3D fluorescence microscopy stack (e.g. confocal or light-sheet microscopy) originating from either animal, fungi or plant systems, stained or expressing a fluorescent reporter highlighting the cell contours (typically, a protein at the plasma membrane). For a better-quality output, it is recommended to use a Z interval of maximum 1 μm." (Procedure section, A. Image Acquisition)
- "if your signal is very heterogeneous, e.g. for cortical microtubules, a higher [Gaussian blur radius] value can help homogenize the signal and obtain a good surface detection." (Procedure section, C. Calibration, step 6.)
- "The voxel properties of your image in micrometers, are automatically filled based on the metadata of the image. If no data is found, these values will all be set to 1." (Procedure section, C. Calibration, step 10.)
- "Remember that the stack should be in .tif and that the top of the stack should also be the top of the sample." (Procedure section, D. Running the script in batch mode, step 18.)

3. Use cases
   - Use case 1: 3D light sheet microscopy images of a *Tribolium* epithelium. We have a 3D stack, .tif format, we know the pixel size, and the life-actin GFP marker signal delimits well a relatively thin epidermal layer. However, these are time-lapse data (SurfCut can process only one time-point at a time), and the tissue is very curvy. The data could be processed by first extracting each individual time points and then analysing the images in a batch after having defined the proper parameters in the calibrate mode. However, SurfCut is not recommended in this case due to the high curvature of the tissue.
   - Use case 2: 3D confocal and spinning disk microscopy images of *Drosophila* epithelia (notum and wing disc), image named "notum2_GFP.tif" We have a 3D stack, in .tif format. We know the pixel size from the description of the Zenodo upload, and the E-Cadherin marker delimits well a relatively thin epidermal layer, which is only slightly curved. SurfCut is appropriate here, since the tissue is not too curvy. Moreover, SurfCut can help remove noise above and below the epidermis and hence render a sharper projection of the cell contours. After adding the pixel size specified in the description of the dataset to the metadata of the image, we can process the image with SurfCut using the following parameters: gaussian blur of radius 3; threshold of 50; top = 6; bottom = 11.

---

16 ▶ https://github.com/sverger/SurfCut/blob/master/SurfCut_UserGuide.pdf.

— Use case 3: 3D light sheet microscopy images of single cells.
We have 3D stacks, .tif format, and we know the pixel size from the AcqInfo.txt file. However, these are 3D closed objects with quite some relief. Here SurfCut is not appropriate to project the 3D stacks, it would deform the cells too much.

— Use case 4: 3D light sheet images of a gastric cancer spheroid.
We have a 3D stack in .tif format. However, the stack contains the first bright and blurry slice that needs to be removed first, the proper voxel size needs to be set based on the information found in the description of the dataset, and the z resolution is rather low (5 micron) compared to the x and y (1 micron). SurfCut can help remove noise around the spheroid, as well as the blur from inside, and render a sharper projection of the surface. We can process the stack with SurfCut using the following parameters: Gaussian blur of radius 3; threshold of 20; top = 0; and bottom = 25.

6

6.4.4 Step 4. Identification of Components in the Code

In the previous steps, we identified the components of the workflow from the text, drew a workflow scheme (see ◘ Fig. 6.4) and checked the prerequisites in terms of data input. If the workflow could be reused as it is, we could have stopped there. Now, we assume that we need to modify the workflow to adapt it to our needs. Hence we need to get a more in-depth knowledge of the code.

Each programming language has a different syntax, but there should always be comments, variables with meaningful names, functions, and other common recognizable items. They can help you understand the structure of the workflow in the code. Read first the comments around the code to identify the different components of the workflow, as found in step 1 and 2 (see ◘ Fig. 6.4). Each component should ideally match with a block of code containing one or several built-in or custom functions, some loops and conditional statements etc. To further understand the order of execution of the workflow, identify also the different input and output variables.

SurfCut contains several defects often found in real codes, and especially in ImageJ macros.[17] We will see, for instance, in the exercise below that, in SurfCut, some components are spread over several blocks of code and intermingled with other components. We will see also that some components in SurfCut are made of several built-in functions that are not wrapped in one bigger function. Of course, a modular code with clearly separated blocks of code and one function per component is easier to read and understand, but SurfCut is representative of ImageJ macros. This lack of structure comes from two elements: (i) macro authors are seldom software developers and hence lack good code writing practices (commenting, wrapping components into functions...) and (ii) most macro authors rely on the macro recorder to find the proper functions to use. The macro recorder prints the macro commands corresponding to the steps done manually by the user through the graphical user interface of ImageJ/Fiji. Whereas some components correspond to a single ImageJ macro built-in function (e.g. a Gaussian blur), other require several functions (e.g. Edge detect). The modular structure with components is lost when using the macro recorder. In addition to these defects, SurfCut contains many repetitions of code lines. This is due

---

17 ▸ https://imagej.nih.gov/ij/developer/macro/macros.html.

to the presence of two types of workflows in one code, the calibrate and the batch workflows, and the lack of optimization in the code to reuse functionalities of one workflow in the other rather than copy functionalities.

As explained in the introduction, we took into account all these defects and carried out a complete refactoring of the code (as described in Step 5 and 6 of this book chapter), fixed bugs, and created new functions to reach a new version of SurfCut, called SurfCut2.[18] We also made a simpler version of the macro, called SurfCut2-Lite. We propose two alternatives to the exercises below, corresponding to two levels of difficulty. For the beginner level, use the SurfCut2-Lite code[19] and do the exercises 4.1, 6, 7 and 8 (skip exercise 4.2 and 5, which are already implemented in the code of SurfCut2-Lite). For the advanced level, use the code of the original SurfCut macro and follow all the steps and exercises proposed.

❓ Exercise 4

Using either SurfCut2-Lite code ("beginner level") or SurfCut code ("advanced level"):

1.   Identify blocks of code corresponding to the different components identified in Step 1 and Step 2.
2.   Extract in a separate text file a minimal version of the macro corresponding to the workflow only: remove user interfaces, "for" loops used to run the batch mode, and "while" loops (in this case, they are not a part of the workflow). Keep only the essential elements the workflow and group elements corresponding to a given component together. Identify the different components of the workflow using the comments present in the macro.

✅ Solution to Exercise 4

"Beginner Level": Response to Task 1, Considering SurfCut2-Lite Code

1.   The workflow appears once, and can be identified at the early part of the macro, in the form of a suite of user-defined functions (line 51–68; similar to the solution of Exercise 5.2). Further, all the components of the workflow are organized as user-defined functions, between line 112 and 222 of the macro (Component 1: line 114–118; Component 2: 120–124; Component 3: 126–131, Component 4: 133–156; Ccomponent 5: 158–207; Component 6: 209–215; similar to exercise 5.1 solution). Note that the Component 5 was split into two user-defined functions (ZAxisShifting and masking), which can be useful and will be explained later in this book chapter.

"Advanced Level": Response to Tasks 1 and 2, Considering the Original SurfCut Code

1.   In the original SurfCut code, the workflow is present twice: Once in the "Calibrate" mode, in which most of the steps are intertwined with user input and interaction, and once in the "Batch" mode, in which the backbone of the macro is embedded in a batch processing "for" loop. The most easily identifiable backbone of the workflow is present between lines 403 and 463 of the macro (Component 1: line 403; Component 2: 404; Component 3: 407–409, Component 4: 418–431; Component

---

18   ▶ https://github.com/VergerLab/SurfCut2.
19   ▶ https://github.com/VergerLab/SurfCut2/blob/master/SurfCut2-Lite.ijm.

5: 433–453; Component 6: 462–463), within the "Batch" mode part of the code. In the "Calibrate" part of the code, equivalent code blocks can be found at lines 68–88 and 161–200.

2. The code below shows a possible solution for the extraction of the minimally required code for the core functionalities of SurfCut. Each component is labeled in a corresponding comment by its corresponding number (see ◻ Fig. 6.4).

```
1    //=Component1=// 8bit conversion
2    run("8-bit");
3
4    //=Component2=// Denoising
5    run("Gaussian Blur...", "sigma=&Rad stack");
6
7    //=Component3=// Binarization
8    setThreshold(0, Thld);
9    run("Convert to Mask", "method=Default background=Light");
10   run("Invert", "stack");
11
12   //=Component4=// Edge detection
13   print (slices);
14   for (img=0; img<slices; img++){
15           print("Edge detect projection" + img + "/" + slices);
16           slice = img+1;
17           selectWindow(list[j]);
18           run("Z Project...", "stop=&slice projection=[Max
         ↪   Intensity]");
19   }
20   print("Concatenate images");
21   run("Images to Stack", "name=Stack title=[]");
22   wait(1000);
23   selectWindow(list[j]);
24   close();
25
26   //=Component5=// Masking
27   //Substraction2
28   print("Substraction2");
29   selectWindow("Stack");
30   run("Duplicate...", "title=Stack-1 duplicate range=1-&slices");
31   open(dir+File.separator+list[j]);
32   wait(1000);
33   run("8-bit");
34   run("Invert", "stack");
35   imageCalculator("Subtract create stack", "Stack-1",list[j]);
36   //Substraction1
37   print("Substraction1");
38   selectWindow("Stack");
39   run("Invert", "stack");
40   getDimensions(w, h, channels, slices, frames);
41   Slice1 = Cut2 +1 - Cut1;
42   Slice2 = slices - Cut1;
43   run("Duplicate...", "title=Stack-2 duplicate
         ↪   range=&Slice1-&Slice2");
44   selectWindow("Result of Stack-1");
45   run("Invert", "stack");
```

```
46    imageCalculator("Subtract create stack", "Stack-2","Result of
      ↪   Stack-1");

47
48    //=Component6=//Z projection
49    print("Project and save SurfCutProj");
50    run("Z Project...", "projection=[Max Intensity]");
```

SurfCutCrudeExtractedWorkflow.ijm

Code available in the GitHub repository of this book.[20]

6.4.5 Step 5. Code Refactoring

In Step 4, we identified the basic components of the workflow in the code and extracted a minimal version of the code. To simplify the later replacement of a component in the code, we propose an optional step: refactoring the code. This step aims at reorganizing the code in order to improve its design and re-usability without changing its input or behavior. The refactored code will be constituted of several user-defined functions, each corresponding to one component of the workflow. The replacement of a component is then equivalent to replacing a function.

Here, we also suggest to split one of the components into two, as a part of the refactoring process. Indeed, while some of the workflow components described in the publication text (and identified in Step 1 and Step 2) correspond to single ImageJ built-in functions, Component 4, "Edge detection", and Component 5, "Masking", with implementation inspired by the algorithm used in the software MorphoGraphX, correspond to many lines of code directly coming from the macro recorder. To improve the organization and re-usability of the code, here we suggest splitting the code corresponding to Component 5, "Masking", in two components. The purpose of Component 5 is to extract a layer of signal in the original stack, using the mask created in the preceding "edge-detection" step. This works by successively shifting the mask down and subtracting the signal twice: once above and once below the signal of interest. So, in fact, it is not only a masking step, but also a Z-axis shifting of the mask preceding the masking. Here, we propose to keep roughly the same process, but reorganize the order in which the steps are taken and separate these two steps: to first create a layer mask by two successive Z-axis shifts of the original mask and subtraction from one-another (Component 5a), and then to do the masking itself (Component 5b).

Overall, such a substantial refactoring costs some time and brainpower, but can strongly improve the workflow and ultimately simplify the replacement of components or their parts, as we will see in the next step.

❓ Exercise 5 ("Advanced Level" Only, Using the Original SurfCut Code)
1. Inspect each component extracted in Step 4, identify unnecessary or disorganized lines of code and optimize the code of each component by simplifying, cutting, and reorganising the code lines.

---

20 ▶ https://github.com/NEUBIAS/neubias-springer-book-2021/blob/master/
Ch07_SurfCut_macro_deconstruction/Exercises_solution_code/SurfCutCrudeExtractedWorkflow.
ijm.

Convert each component into a user-defined function.[21]

As discussed above, re-implement Component 5 of the workflow into two steps: (5a) Layer mask creation, and (5b) Raw signal masking.

2. Write, or extract from the original SurfCut macro, the lines corresponding to the definition of the input and the parameters necessary to run the code.

Make a working macro including the definition of parameters at the beginning and the fully refactored version of the workflow (optimized code, user-defined functions, and with Component 5 split in two parts).

✅ **Solution to Exercise 5**

1. A possible refactoring of the initial code into functions, with the re-implementation of Component 5 in two parts, is shown below. Some of the variable names have been homogenized, some unnecessary code lines (e.g. wait(1000);) have been removed, and all the components have been transformed into simple user-defined functions. To get a better insight, compare the code proposed below with the equivalent code extracted from the SurfCut macro in Step 4.

```
33   //=Component1=//
34   function BitConversion(){
35           print ("Pre-processing");
36           run("8-bit");
37   };
38
39   //=Component2=//
40   function Denoising(Rad){
41           //Gaussian blur (uses the variable "Rad" to define the
                 ↪  sigma of the gaussian blur)
42           print ("Gaussian Blur");
43           run("Gaussian Blur...", "sigma=&Rad stack");
44   };
45
46   //=Component3=//
47   function Binarization(Thld){
48           //Object segmentation (uses the variable Thld to define
                 ↪  the threshold applied)
49           print ("Threshold segmentation");
50           setThreshold(0, Thld);
51           run("Convert to Mask", "method=Default
                 ↪  background=Light");
52   };
53
54   //=Component4=//
55   function EdgeDetection(imgName){
56           print ("Edge detect");
57           //Get the dimensions of the image to know the number of
                 ↪  slices in the stack and thus the number of loops to
                 ↪  perform
58           getDimensions(w, h, channels, slices, frames);
59           print (slices);
60           run("Invert", "stack");
```

---

21 ▶ https://imagej.nih.gov/ij/developer/macro/macros.html.

```
61    for (img=0; img<slices; img++){
62            //Display progression in the log
63            print("Edge detect projection" + img + "/" +
              ↪  slices);
64            slice = img+1;
65            selectWindow(imgName);
66            //Successively projects stacks with increasing
              ↪  slice range (1-1, 1-2, 1-3, 1-4,...)
67            run("Z Project...", "stop=&slice
              ↪  projection=[Max Intensity]");
68    };
69    //Make a new stack from all the Z-projected images
      ↪  generated in the loop above
70    run("Images to Stack", "name=Mask title=[]");
71    selectWindow(imgName);
72    close();
73    //Close binarized image generated in component2
      ↪  (imgName), but keeps the image (Mask) generated
      ↪  after the edge detect.
74 };
75
76 //=Component5a=//
77 function ZAxisShifting(Cut1, Cut2){
78        print ("Layer mask creation");
79        ///First Z-axis shift
80        //Get dimension w and h, and pre-defined variable Cut1
          ↪  depth to create an new "empty" stack
81        getDimensions(w, h, channels, slices, frames);
82        newImage("Add1", "8-bit white", w, h, Cut1);
83        //Duplicate and invert Mask while removing bottom
          ↪  slices corresponding to the Z-axis shift (Cut1
          ↪  depth)
84        Slice1 = slices - Cut1;
85        selectWindow("Mask");
86        run("Duplicate...", "title=Mask1Sub duplicate
          ↪  range=1-&Slice1");
87        run("Invert", "stack");
88        //Add newly created empty slices (Add1) at begining of
          ↪  Mask1Sub, thus recreating a stack with the original
          ↪  dimensions of the image and in whcih the mask is
          ↪  shifted in the Z-axis.
89        run("Concatenate...", " title=[Mask1] image1=[Add1]
          ↪  image2=[Mask1Sub] image3=[-- None --]");
90        ///Second Z-axis shift
91        //Use image dimension w and h from component3 and
          ↪  pre-defined variable Cut2 depth to create an new
          ↪  "empty" stack
92        newImage("Add2", "8-bit black", w, h, Cut2);
93        //Duplicate Mask while removing bottom slices
          ↪  corresponding to the Z-axis shift (Cut2 depth)
94        Slice2 = slices - Cut2;
95        selectWindow("Mask");
96        run("Duplicate...", "title=Mask2Sub duplicate
          ↪  range=1-&Slice2");
97        //Add newly created empty slices (Add2) at begining of
          ↪  Mask2Sub,
```

```
98          run("Concatenate...", "  title=[mask2] image1=[Add2]
            ↪   image2=[Mask2Sub] image3=[-- None --]");
99          //Subtract both shifted masks to create a layer mask
100         imageCalculator("Add create stack", "Mask1","mask2");
101         close("Mask");
102         close("Mask1");
103         close("Mask2");
104         selectWindow("Result of Mask1");
105         rename("LayerMask");
106         //Close original and shifted masks ("Mask", "Mask1" and
            ↪   "Mask2"), but keeps the newly created "layerMask"
            ↪   resulting from the subtraction of the two shifted
            ↪   masks.
107    };
108
109    //=Component5b=//
110    function Masking(imgPath, imgName){
111         print ("Cropping stack");
112         //Open raw image
113         open(imgPath);
114         run("Grays");
115         //Apply LayerMask to raw image
116         imageCalculator("Subtract create stack", imgName,
            ↪   "LayerMask");
117         close("LayerMask");
118    };
119
120    //=Component6=//
121    function ZProjections(imgName){
122         selectWindow("Result of " + imgName);
123         run("Z Project...", "projection=[Max Intensity]");
124         rename("SurfCut projection");
125         selectWindow(imgName);
126         run("Z Project...", "projection=[Max Intensity]");
127         rename("Original projection");
128    };
```
SurfCutWorkflowFunc.ijm

Code available in the GitHub repository of this book.[22]

2. The code for opening an image and getting the variable names can be found in lines 32–38 of the original SurfCut macro. Variables related to the radius of the Gaussian blur filter, threshold for the segmentation, and top and bottom depths (in micron) for masking can be found in lines 42–45 and 149–150. These variables are used for the definition of Cut1 and Cut2 (the actual values of the Z-axis shifts, depending on the thickness of the stack slice steps). The values of Cut1 and Cut2 depend on the stack slice thickness, which, in the macro, is extracted from the metadata of the image (line 96). For simplicity, we can here define it directly in the code.

---

22 ▶ https://github.com/NEUBIAS/neubias-springer-book-2021/blob/master/
Ch06_SurfCut_macro_deconstruction/Exercises_solution_code/SurfCutWorkflowFunc.ijm.

In the solution below, the parameters identified above are called at the beginning of the macro. The functions defined in the answer to Question 1 (above) are successively called, giving a clear overview and easy reading of the workflow.

```
1   ///Parameters
2   Rad = 3;
3   Th1d = 20;
4   Top = 6;
5   Bot = 8;
6   Dpt = 0.5;
7   Cut1= Top/Dpt;
8   Cut2= Bot/Dpt;
9
10  ///Open a stack and get names
11  open();
12  imgDir = File.directory;
13  print(imgDir);
14  imgName = getTitle();
15  print(imgName);
16  imgPath = imgDir+imgName;
17  print(imgPath);
18
19  ///SurfCut Workflow User-Defined Functions
20  BitConversion(); //Component1
21  Denoising(Rad); //Component2
22  Binarization(Th1d); //Component3
23  EdgeDetection(imgName); //Component4
24  ZAxisShifting(Cut1, Cut2); //Component5a
25  Masking(imgPath, imgName); //Component5b
26  ZProjections(imgName); //Component6
27
28  ///End
29  print("=== Done ===");
```
SurfCutWorkflowFunc.ijm

Code available in the GitHub repository of this book.[23]

6.4.6 Step 6. Replacing a Component: Shift Mask in the Z-Axis Direction

Now we have a well organized and flexible workflow. It is time to inspect it in detail and determine if the different components are best adapted to our needs. As an example, we will now examine the Component 5a "Layer mask creation" created during the refactoring of the SurfCut macro code (line 76–107); this component is included in the SurfCut2-Lite code (line 158–197). In the current implementation, we use a sequential Z-axis shift of the mask to make a layer-mask. But in principle, as discussed in (Erguvan et al., 2019), a 3D erosion, although more computationally demanding, would be more suitable for samples with high curvature. Let us try to

---

23 ▶ https://github.com/NEUBIAS/neubias-springer-book-2021/blob/master/
Ch07_SurfCut_macro_deconstruction/Exercises_solution_code/SurfCutWorkflowFunc.ijm.

replace the Component 5a with a procedure which uses a 3D erosion instead of a Z-axis shift.

❓ Exercise 6

1. Find how to perform a 3D erosion on a binary object using Fiji.
2. Write a function similar to the existing Component 5a, but performing a 3D erosion on the mask, instead of Z-axis shift (to be used to replace Component 5a).
3. Modify the refactored version of SurfCut or SurfCut2-Lite to include both alternatives for processing (Z-axis shift and erosion) with a conditional statement.

✅ Solution to Exercise 6

1. 3D erosion operation is available in Fiji Plugins>Process>Erode(3D). The macro recorder can be used to record the corresponding code. Alternatively, we can apply the erosion operation from the 3D suite plugin (Ollion et al., 2013).[24]
2. Below we propose a user-defined function that takes as input parameter two "erosion depths", i.e. two distances (in pixels or microns) from the surface, defining the upper and lower boundary of the signal to be extracted. This function processes the binary stack ("Mask") obtained with the EdgeDetect. The mask is eroded by several erosion steps using a "for" loop. The number of steps depends on the value of the first erosion depth. The image resulting from this first erosion is then duplicated and eroded further to reach the second defined value of erosion depth. The first eroded stack is then inverted (in terms of binary values), and these two eroded stacks (binary values) are summed, forming a layer mask ("LayerMask") that can then be used in Component 5b.

```
116  //=Component5a=//
117  function Erosion(Ero1, Ero2){
118          print ("Layer mask creation - Erosion");
119          //Erosion 1
120          selectWindow("Mask");
121          run("Duplicate...", "title=Mask-Ero1 duplicate");
122          print("Erosion1");
123          print(Ero1 + " erosion steps");
124          for (erode1=0; erode1<Ero1; erode1++){
125                  print("Erode1");
126                  run("Erode (3D)", "iso=255");
127          };
128          //Erosion 2 (here instead of restarting from the
             ↪   original mask, the eroded mask is duplictaed and
             ↪   further eroded. In this case Ero2 corresponds
129          //to the number of additional steps of erosion, or the
             ↪   thickness of the future layer mask)
130          selectWindow("Mask-Ero1");
131          run("Duplicate...", "title=Mask-Ero2 duplicate");
132          print("Erosion2");
133          print(Ero2 + " erosion steps");
```

24 ▶ https://imagej.net/3D_ImageJ_Suite.

```
134          for (erode2=0; erode2<Ero2; erode2++){
135                  print("Erode2");
136                  run("Erode (3D)", "iso=255");
137          };
138          selectWindow("Mask-Ero1");
139          run("Invert", "stack");
140          //Subtract both shifted masks to create a layer mask
141          imageCalculator("Add create stack",
                ↪   "Mask-Ero1","Mask-Ero2");
142          close("Mask");
143          close("Mask-Ero1");
144          close("Mask-Ero2");
145          selectWindow("Result of Mask-Ero1");
146          rename("LayerMask");
147          //Close original and eroded masks ("Mask", "Mask-Ero1"
                ↪   and "Mask-Ero2"), but keeps the newly created
                ↪   "layerMask" resulting from the subtraction of the
                ↪   two eroded masks.
148    };
```

SurfCutWorkflowFuncErode.ijm

Code available in the GitHub repository of this book.[25]

3. In the example below, we added the new component 5a as a function. We either call the Z-shift or erode function, using a conditional "if" and "else if" statement (lines 27–31 below). In addition, we defined new variables necessary for the new erode function and for the conditional statement (lines 9–11): Ero1 and Ero2 which are calculated from the values Cut1 and Cut2, and MODE, in which the user can define whether to process the macro with the Z-axis shift, or using the erosion.

```
1    ///Parameters
2    Rad = 3;
3    Thld = 20;
4    Top = 6;
5    Bot = 8;
6    Dpt = 0.5;
7    Cut1= Top/Dpt;
8    Cut2= Bot/Dpt;
9    Ero1 = Cut1;
10   Ero2 = Cut2-Cut1;
11   MODE = "erode"; //(or "Z-shift")
12
13   ///Open a stack and get names
14   open();
15   imgDir = File.directory;
16   print(imgDir);
17   imgName = getTitle();
18   print(imgName);
```

25 ▶ https://github.com/NEUBIAS/neubias-springer-book-2021/blob/master/
Ch07_SurfCut_macro_deconstruction/Exercises_solution_code/SurfCutWorkflowFuncErode.
ijm.

```
19   imgPath = imgDir+imgName;
20   print(imgPath);
21
22   ///SurfCut Workflow User-Difined Functions
23   BitConversion(); //Component1
24   Denoising(Rad); //Component2
25   Binarization(Thld); //Component3
26   EdgeDetection(imgName); //Component4
27   if (MODE=="erode"){ //Component5a
28           Erosion(Ero1, Ero2);
29   } else if (MODE=="Z-shift"){
30           ZAxisShifting(Cut1, Cut2);
31   };
32   Masking(imgPath, imgName); //Component5b
33   ZProjections(imgName); //Component6
34
35   ///End
36   print("=== Done ===");
```

SurfCutWorkflowFuncErode.ijm

Code available in the GitHub repository of this book.[26]

6.4.7 Step 7. Benchmarking: Comparison of Two Alternative Components

Benchmarking is assessment of benefits and drawbacks of different algorithms and evaluation of their performance in terms of a range of criteria: speed, memory usage when dealing with, e.g., 2D or 3D stacks, and quality of the result (e.g., How much the projection deforms the image? How well does the selected filter extract the features of interest in the image?). Benchmarking can be performed for two (or more) similar components, or two (or more) similar workflows. Here, we would like to assess if the change of a component that we made in Step 6 is beneficial for the workflow.

In Step 6, we suggested an alternative code for layer mask creation. We will now benchmark this new workflow against the original one. First, we can look at the output and qualitatively assess if the workflow generates the expected result. Second, and importantly—we will quantitatively assess the impact of the change on the performances of the workflow, mainly evaluating if the processing time is a limiting factor. This can be done very quickly by adding timestamps in the script, and calculating the time elapsed between the beginning and the end of the workflow execution. Furthermore, using nested "for" loops, it is also possible to iteratively test how different parameters affect the workflow processing time: Erode or Z-shift, and increasing depth of masking. The values can also be recorded in a text file to make a direct comparison of the performance after running the benchmark.

---

26 ▶ https://github.com/NEUBIAS/neubias-springer-book-2021/blob/master/
Ch07_SurfCut_macro_deconstruction/Exercises_solution_code/SurfCutWorkflowFuncErode.
ijm.

✅ Exercise 7

1. Find out how to add a timestamp in the macro.
2. Implement a way to quantify the processing time of the macro.
3. Implement a way to record the processing times in a text file.
4. Implement nested "for" loops to iteratively test the Erode and the Z-shift, as well as the increasing depth of masking, starting from 1 (Top) and 2 (Bot), and reaching 5 (Top) and 6 (Bot) (1–2, 2–3, 3–4, 4–5, 5–6). Furthermore, a simple way to decrease processing time in ImageJ macros is to use the "setBatchMode" function. It allows the processing of the images to be carried out without displaying the images, which can improve processing time by up to a factor of 20. Implement an additional nested loop to test how much the "setBatchMode" function improves the performances of the macro.
5. Run this modified version of SurfCut on the provided SurfCut data (see ▶ Section 6.2) and compare the performances of the two components.

✅ Solution to Exercise 7

1. Within the ImageJ built-in functions, there are at least two ways to add a time-stamp: "GetDateAndTime" and "GetTime". The latter is more practical to calculate the elapsed time, because it gives a value in milliseconds, instead of hours:minutes:seconds:milliseconds (which is less practical for further analysis).
2. "GetTime" can be added right before and right after the execution of the workflow of interest. Subtracting the value given at the first time point from the value obtained at the second time point gives the elapsed time.

```
51          T0 = getTime();
     [Workflow]
67          T1 = getTime();
68          T=T1-T0;
69          print(T + "msec");
```

SurfCutWorkflowBenchmark.ijm

3. A text file can be created with the built-in function "File.open", it can be closed using the function "File.close", and text can be added to the closed file with the function File.append. Note that, while the text can also be written directly in an open text file with the "print" function, only one file can be opened at a time, which can be limiting in some situations (e.g. if other parameters are being recorded by the macro in another text file).

    Create file (with Headers):

```
13   f = File.open(imgDir + File.separator + "MultiBenchmark.txt");
14   print(f, "Mode\tBatch\tTop\tBot\tTime(msec)");
15   File.close(f);
```

Append file with recorded data:

```
70          File.append(MODE + "\t"+ BATCH + "\t" + Top + "\t" + Bot
     ↳      + "\t" + T, imgDir + File.separator +
     ↳      "MultiBenchmark.txt");
```

SurfCutWorkflowBenchmark.ijm

4. Three nested "for" loops need to be implemented to iteratively test the three types of parameters of interest. This requires to slightly reorganise where the variables are defined, since a number of them are now defined in the "for" loops. A text file

is saved containing the output of all the time elapse measurements along with the parameter used at each iteration. The SurfCut 2D projection is also saved in each case, in order to assess the quality of the output. A possible complete solution is shown below.

```
1    ///Open a stack and get names
2    open();
3    imgDir = File.directory;
4    print(imgDir);
5    imgName = getTitle();
6    print(imgName);
7    imgPath = imgDir+imgName;
8    print(imgPath);
9    selectWindow(imgName);
10   close();
11
12   //Make tab separated file to record the benchmarking data
13   f = File.open(imgDir + File.separator + "MultiBenchmark.txt");
14   print(f, "Mode\tBatch\tTop\tBot\tTime(msec)");
15   File.close(f);
16
17   //Nested "for" loops
18   //Loop parameters
19   mode = newArray("Z-Shift", "erode");
20   batch = newArray(true, false);
21   TopDepth = 5;
22
23   //Nested loops
24   //loop between Z-shift and erode
25   for (Mode = 0; Mode<mode.length; Mode++){
26     //loop between "setBatchMode" true and false
27     for (Batch = 0; Batch<batch.length; Batch++){
28       //loop through increasing depths for cutting
29       for (Top = 1; Top < TopDepth; Top++){
30
31         ///Parameters
32         Rad = 3;
33         Thld = 20;
34         Bot = Top+1; //Automatically make mask layer thickness to
              ↪   1 micron
35         Dpt = 0.5;
36         Cut1= Top/Dpt;
37         Cut2= Bot/Dpt;
38         Ero1 = Cut1;
39         Ero2 = Cut2-Cut1;
40         MODE = mode[Mode];
41         BATCH = batch[Batch];
42
43         print("Mode : " + MODE + " Batch : " + BATCH + " Top = "
              ↪   + Top + " Bot = " + Bot);
44
45         setBatchMode(BATCH);
46
47         //Open predefined image for precessing in the loop
48         open(imgPath);
49
50         //Benchmark T0
51         T0 = getTime();
```

```
52
53          ///SurfCut Workflow User-Defined Functions
54          BitConversion(); //Component1
55          Denoising(Rad); //Component2
56          Binarization(Thld); //Component3
57          EdgeDetection(imgName); //Component4
58          if (MODE=="erode"){ //Component5a
59            Erosion(Ero1, Ero2);
60          } else if (MODE=="Z-Shift"){
61            ZAxisShifting(Cut1, Cut2);
62          };
63          Masking(imgPath, imgName); //Component5b
64          ZProjections(imgName); //Component6
65
66          //Benchmark T1
67          T1 = getTime();
68          T=T1-T0;
69          print(T + "msec");
70          File.append(MODE + "\t"+ BATCH + "\t" + Top + "\t" + Bot
            ↪   + "\t" + T, imgDir + File.separator +
            ↪   "MultiBenchmark.txt");
71
72          //Save SurfCut output
73          selectWindow("SurfCut projection");
74          saveAs("Tiff", imgDir + File.separator +
            ↪   "SurfCutBenchmark_mode-"+ MODE + "_Batch-"+ BATCH +
            ↪   "_Top-" + Top + "_Bot-" + Bot + ".tif");
75
76          run("Close All");
77
78          //End of nested loops
79        };
80      };
81    };
82
83    ///End
84    print("=== Done ===");
```
SurfCutWorkflowBenchmark.ijm

Code available in the GitHub repository of this book.[27]

5. With the Erode component, the processing time increases linearly with the number of erosion steps required. This is not the case with the Z-axis shift component. However, the accuracy and quality of the result for samples with higher curvature are in principle improved with the Erode process, as described in Erguvan et al., 2019.

---

27 ▶ https://github.com/NEUBIAS/neubias-springer-book-2021/blob/master/Ch07_SurfCut_macro_deconstruction/Exercises_solution_code/SurfCutWorkflowBenchmark.ijm.

6.4.8 Step 8. Linking to Another Workflow: FibrilTool

As described in the introduction and the reference publication, SurfCut was designed, and has been used, as a pre-processing step for cell segmentation and cortical microtubule (CMT) signal analysis with another ImageJ macro called FibrilTool (Boudaoud et al., 2014). In this case, SurfCut and FibrilTool are two components of a workflow. However, the output of SurfCut cannot be directly taken as the input of FibrilTool. As a final exercise, we will analyse how these two components can be linked in a workflow, by adding one, or several, intermediate components.

> **? Exercise 8**
> 1. Identify the type of input required for the FibrilTool macro.
> 2. Identify the missing steps between the SurfCut output and the FibrilTool input, required to connect the two components.
> 3. How would you implement these missing steps?
> 4. Last but not least, consider whether the required tools already exist, or you need to implement them *de novo*.

> **✓ Solution to Exercise 8**
> 1. FibrilTool takes as input an ImageJ ROI and a corresponding image containing the fibrilar structure to analyse (e.g. CMTs).
> 2. Surfcut can directly generate one of the FibriTool inputs: the preprocessed CMT image. It can also generate the cell contour image. The missing step here is the generation of ROIs from this cell contour image. Finally, the originally published version of FibrilTool takes and analyses ROIs manually one by one. Since many ROIs per image can be created, FibrilTool could be automatized to analyse all these ROIs automatically, one after the other.
> 3. For generation of ROIs from a cell contour image, a simple Analyze particle function may be sufficient. However, to ensure better results, a watershed segmentation could be used. For FibrilTool automation, a "for" loop can be implemented to analyse automatically all ROIs generated in the preceding step.
> 4. The second part of the user guide of SurfCut describes these additional workflow components. Previously, an automated version of Fibriltool that uses ROIset.zip as input, instead of individual ROIs, was implemented: "FibrilToolBatch.ijm" (Louveaux and Boudaoud, 2018). We then implemented a macro called "segmentation4FTBatch.ijm".[28] This macro uses the MorpholibJ morphological segmentation tool (Arganda-Carreras et al., 2020; Legland et al., 2016) to segment the cell contours extracted from SurfCut, and other ImageJ functions to ultimately generate a ROIset.zip used as input for FibrilToolBatch.

6.5 Analysis of the Results: Presentation and Discussion

In this chapter, we performed the deconstruction of the ImageJ macro SurfCut in eight steps. In Step 1, we identified 6 main components: 8bit conversion, Gaussian blur

---

28 ▶ https://github.com/sverger/Segmentation4FTBatch.

denoising, threshold binarization, "edge detect", signal masking, and Z-projection, by reading the available description of the workflow (Erguvan et al., 2019). We also learned that the macro has (i) a single processing mode ("Calibrate"), with many user interactions and (ii) a "batch" mode.

In Step 2, we inferred a first workflow scheme, using the findings of Step 1.

In Step 3, we went back to reading the textual description, in order to identify which type of data can be used as input to the workflow. We found that 3D confocal stacks of a moderately curved tissue were the characteristic type of input of this macro.

In Step 4 we went through the code in the macro and identified the ~50 lines of codes that compose the backbone of the workflow. We found that the components were much more interwoven in the code than in the corresponding text description. We also found that some of the workflow components (as described in the available macro description) correspond to roughly a single ImageJ built-in function, while others are custom multi-line implementations of processes within the macro.

In Step 5, we cleaned the code by removing all the batch-loops, user interactions and accessory code lines, and separated all of the identified components into user-defined functions. While being optional, the clear separation of components in code blocks provided a much more readable and re-usable version of the macro, that could be further modified without the risk of breaking the whole workflow.

In Step 6, after deeply deconstructing and refactoring the macro, we replaced one of the critical steps of the workflow. We identified how to interfere with the SurfCut initial process, and we replaced the Z-axis shift of the mask by multiple steps of 3D erosion.

In Step 7, we benchmarked this new implementation and revealed that erosion, in principle, provides a more accurate extraction of layer signal, especially for curved samples; however the processing time increases linearly with the number of erosion steps required. This is not the case for the Z-axis shift.

Finally, in Step 8, we explored the possibility to embed SurfCut in a larger workflow, and we took the example of a combination with another macro called FibrilTool. We identified a missing component to link both workflows: the creation of regions of interest (ROIs) from the segmentation of the cell contours generated by SurfCut.

6.6 Concluding Remarks

We think that an important part of the work of a bioimage analyst is assessment of the relevance of a published workflow, and—if suitable—its adaptation and optimization to own needs. Such an approach, instead of coding everything from scratch, can save a lot of time. In this chapter, we proposed a generic way to deconstruct a workflow published in a scientific paper. The deconstruction was performed in eight steps, starting with reading the paper, and reaching inspection and modification of the code. Not only this can help to gain time by avoiding to "reinvent the wheel", but, in our opinion, reviewing and modifying the code of someone else helps reflecting on one's own code and coding practices.

We chose a representative example of a workflow rather than an ideal case, to underline the challenges that the deconstruction can bring. Mainly, a poorly organised code can make identification of the components and their replacement challenging. To address the issue, we suggested an optional re-factorisation step. Reorganising the code into separate blocks, or functions corresponding to components, is optional but has many benefits and should not be underestimated. First, by simplifying the structure of the code, components are easier to identify and replace. Second, this practice leads to better understanding of the workflow. Third, it minimizes the risk of introducing errors. In our opinion, one should always weigh the pros and the cons of refactoring a code, before modifying it. We also introduced a benchmarking step to insist on the fact that the benefits of workflow modifications should be assessed, and modified workflows published along with some explanations and justifications of the changes made.

> **Take-Home Message**
>
> Deconstructing a workflow written and designed by someone else can be a challenging task. In this chapter, through successive steps, we propose one possible approach to this problem. By looking at the available description of the workflow (step 1), drawing a workflow scheme identifying the different components (step 2) and assessing the prerequisites and limitations in terms of input data (step 3), we can get good initial understanding of what the workflow does, and if it is suitable for the problem we are trying to solve. We can then start to inspect the code. By identifying the basic components of the workflow in the code (step 4), if necessary, refactoring the code (step 5) to make it more readable, reusable and easier to modify, we can get an in-depth knowledge of the workflow and its code.
>
> We can then use and adapt the code to our needs, by replacing or modifying one or several of the components of the workflow (step 6) and assessing if this was beneficial by benchmarking (step 7). In addition, we have gained sufficient information on the studied workflow to be able to link it, or its parts, with other existing workflows (step 8).

Acknowledgements   We would like to thank Kota Miura for inviting us to write this book chapter and reviewing it, Mafalda Sousa for reviewing the chapter and providing data for the Exercise number 3, all the trainees of the NEUBIAS training school 15 who worked on the deconstruction of the SurfCut macro and gave us the idea to write this chapter, our co-authors on the SurfCut publication Özer Erguvan and Olivier Hamant, without whom this work would not exist, and Robert Haase, Léo Valon, Ralitza Staneva, and Meghan Driscoll for publicly sharing datasets that could be taken as example in the Exercise number 3.

References

Arganda-Carreras I, Legland D, Rueden C, Mikushin D, Eglinger J, Burri O, Schindelin J, Helfrich S, Fiedler CC (2020) ijpb/MorphoLibJ: MorphoLibJ 1.4.2.1. https://doi.org/10.5281/zenodo.3826337

Band LR, Wells DM, Fozard JA, Ghetiu T, French AP, Pound MP, Wilson MH, Yu L, Li W, Hijazi HI, Oh J, Pearce SP, Perez-Amador MA, Yun J, Kramer E, Alonso JM, Godin C, Vernoux T, Hodgman TC, Pridmore TP, Swarup R, King JR, Bennett MJ (2014) Systems Analysis of Auxin Transport in the Arabidopsis Root Apex. Plant Cell 26(3):862–875. https://doi.org/10.1105/tpc.113.119495

Baral A, Aryal B, Jonsson K, Morris E, Demes E, Takatani S, Verger S, Xu T, Bennett M, Hamant O, Bhalerao RP (2021) External mechanical cues reveal a katanin-independent mechanism behind auxin-mediated tissue bending in plants. Dev Cell 56(1):67–80.e3, https://doi.org/10.1016/j.devcel.2020.12.008. https://linkinghub.elsevier.com/retrieve/pii/S1534580720309837

Barbier de Reuille P, Bohn-Courseau I, Godin C, Traas J (2005) A protocol to analyse cellular dynamics during plant development: a protocol to analyse cellular dynamics. Plant J 44(6):1045–1053. https://doi.org/10.1111/j.1365-313X.2005.02576.x

Barbier de Reuille P, Routier-Kierzkowska AL, Kierzkowski D, Bassel GW, Schüpbach T, Tauriello G, Bajpai N, Strauss S, Weber A, Kiss A, Burian A, Hofhuis H, Sapala A, Lipowczan M, Heimlicher MB, Robinson S, Bayer EM, Basler K, Koumoutsakos P, Roeder AHK, Aegerter-Wilmsen T, Nakayama N, Tsiantis M, Hay A, Kwiatkowska D, Xenarios J, Kuhlemeier C, Smith RS (2015) MorphoGraphX: a platform for quantifying morphogenesis in 4D. eLife 4:e05864. https://doi.org/10.7554/eLife.05864

Boudaoud A, Burian A, Borowska-Wykret D, Uyttewaal M, Wrzalik R, Kwiatkowska D, Hamant O (2014) FibrilTool, an ImageJ plug-in to quantify fibrillar structures in raw microscopy images. Nat Protoc 9(2):457–463. https://doi.org/10.1038/nprot.2014.024

Candeo A, Sana I, Ferrari E, Maiuri L, D'Andrea C, Valentini G, Bassi A (2016) Virtual unfolding of light sheet fluorescence microscopy dataset for quantitative analysis of the mouse intestine. J Biomed Optics 21(05):1. https://doi.org/10.1117/1.JBO.21.5.056001. https://www.spiedigitallibrary.org/journals/journal-of-biomedical-optics/volume-21/issue-05/056001/Virtual-unfolding-of-light-sheet-fluorescence-microscopy-dataset-for-quantitative/10.1117/1.JBO.21.5.056001.full

Driscoll MK, Welf ES, Jamieson AR, Dean KM, Isogai T, Fiolka R, Danuser G (2019) Robust and automated detection of subcellular morphological motifs in 3D microscopy images. Nat Methods 16(10):1037–1044. https://doi.org/10.1038/s41592-019-0539-z

Erguvan O, Verger S (2019) Dataset of confocal microscopy stacks from plant samples—ImageJ SurfCut: a user-friendly, high- throughput pipeline for extracting cell contours from 3D confocal stacks. BMC Biol 17:38. https://doi.org/10.5281/zenodo.2577053

Erguvan O, Louveaux M, Hamant O, Verger S (2019) ImageJ SurfCut: a user-friendly pipeline for high-throughput extraction of cell contours from 3D image stacks. BMC Biol 17(1):38. https://doi.org/10.1186/s12915-019-0657-1

Galea GL, Nychyk O, Mole MA, Moulding D, Savery D, Nikolopoulou E, Henderson DJ, Greene NDE, Copp AJ (2018) Vangl2 disruption alters the biomechanics of late spinal neurulation leading to spina bifida in mouse embryos. Dis Model Mech 1(3):dmm032219. https://doi.org/10.1242/dmm.032219

Haase R, Royer LA, Steinbach P, Schmidt D, Dibrov A, Schmidt U, Weigert M, Maghelli N, Tomancak P, Jug F, Myers EW (2020) CLIJ: GPU-accelerated image processing for everyone. Nature Methods 17(1):5–6. https://doi.org/10.1038/s41592-019-0650-1

Heemskerk I, Streichan SJ (2015) Tissue cartography: compressing bio-image data by dimensional reduction. Nature Methods 12(12):1139–1142. https://doi.org/10.1038/nmeth.3648

Legland D, Arganda-Carreras I, Andrey P (2016) MorphoLibJ: integrated library and plugins for mathematical morphology with ImageJ. Bioinformatics 32(22):3532–3534. https://doi.org/10.1093/bioinformatics/btw413

Li K, Wu X, Chen D, Sonka M (2006) Optimal surface segmentation in volumetric images—a graph-theoretic approach. IEEE Trans Pattern Anal Mach Intell 28(1):119–134. https://doi.org/10.1109/TPAMI.2006.19

Louveaux M, Boudaoud A (2018) FibrilTool Batch: an automated version of the ImageJ/Fiji plugin FibrilTool. https://doi.org/10.5281/zenodo.2528872

Miura K, Tosi S (2016) Introduction. Wiley-VCH, Weinheim, pp 1–3

Miura K, Tosi S (2017) Epilogue: a framework for bioimage analysis. Wiley, London, p 269–284. https://doi.org/10.1002/9781119096948.ch11

Miura K, Paul-Gilloteaux P, Tosi S, Colombelli J (2020) Workflows and components of bioimage analysis. Springer, Berlin, p 1–7. Learning Materials in Biosciences. https://doi.org/10.1007/978-3-030-22386-1_1

Möller B, Poeschl Y, Plötner R, Bürstenbinder K (2017) PaCeQuant: a tool for high-throughput quantification of pavement cell shape characteristics. Plant Physiol 175(3):998–1017. https://doi.org/10.1104/pp.17.00961

Ollion J, Cochennec J, Loll F, Escudé C, Boudier T (2013) TANGO: a generic tool for high-throughput 3D image analysis for studying nuclear organization. Bioinformatics 29(14):1840–1841. https://doi.org/10.1093/bioinformatics/btt276

Rocha S, Carvalho J, Oliveira C (2020) Gastric cancer spheroid. http://doi.org/10.5281/zenodo.4244952

Schindelin J, Arganda-Carreras I, Frise E, Kaynig V, Longair M, Pietzsch T, Preibisch S, Rueden C, Saalfeld S, Schmid B, Tinevez JY, White DJ, Hartenstein V, Eliceiri K, Tomancak P, Cardona A (2012) Fiji: an open-source platform for biological-image analysis. Nat Methods 9(7):676–682. http://doi.org/10.1038/nmeth.2019

Schmid B, Shah G, Scherf N, Weber M, Thierbach K, Campos CP, Roeder I, Aanstad P, Huisken J (2013) High-speed panoramic light-sheet microscopy reveals global endodermal cell dynamics. Nat Commun 4(1):2207. http://doi.org/10.1038/ncomms3207

Shihavuddin A, Basu S, Rexhepaj E, Delestro F, Menezes N, Sigoillot SM, Del Nery E, Selimi F, Spassky N, Genovesio A (2017) Smooth 2D manifold extraction from 3D image stack. Nat Commun 8(1):15554. http://doi.org/10.1038/ncomms15554

Sánchez-Corrales YE, Hartley M, Van Rooij J, Marée AF, Grieneisen VA (2018) Morphometrics of complex cell shapes: lobe contribution elliptic Fourier analysis (LOCO-EFA). Development 145(6):dev156778. http://doi.org/10.1242/dev.156778

Takatani S, Verger S, Okamoto T, Takahashi T, Hamant O, Motose H (2020) Microtubule response to tensile stress is curbed by NEK6 to buffer growth variation in the arabidopsis hypocotyl. Curr Biol 30(8):1491–1503.e2. https://doi.org/10.1016/j.cub.2020.02.024

Valon L, Staneva R (2020) Dataset of examples of Drosophila epithelia at different developmental stages. https://doi.org/10.5281/zenodo.4114074

Verger S, Long Y, Boudaoud A, Hamant O (2018) A tension-adhesion feedback loop in plant epidermis. eLife 7:e34460. https://doi.org/10.7554/eLife.34460

Viktorinová I, Haase R, Pietzsch T, Henry I, Tomancak P (2019) Analysis of actomyosin dynamics at local cellular and tissue scales using time-lapse movies of cultured drosophila egg chambers. J Vis Exp (148):e58587. https://doi.org/10.3791/58587. https://www.jove.com/video/58587/analysis-actomyosin-dynamics-at-local-cellular-tissue-scales-using

Vorkel D, Haase R, Myers E (2020) Strausberg_tribolium_la-GFP_tailpole_run (Excerpt timepoints 291–340). https://doi.org/10.5281/zenodo.3981193

Wada H, Hayashi S (2020) Net, skin and flatten, ImageJ plugin tool for extracting surface profiles from curved 3D objects. Micropublication Biol p 3. https://doi.org/10.17912/micropub.biology.000292

Wu TC, Belteton S, Pack J, Szymanski DB, Umulis D (2016) LobeFinder: a convex hull-based method for quantitative boundary analyses of lobed plant cells. Plant Physiol 171(4):2331–2342. https://doi.org/10.1104/pp.15.00972

Zubairova US, Verman PY, Oshchepkova PA, Elsukova AS, Doroshkov AV (2019) LSM-W2: laser scanning microscopy worker for wheat leaf surface morphology. BMC Syst Biol 13(S1):22. https://doi.org/10.1186/s12918-019-0689-8

i.2.i. with the (Fruit) Fly: Quantifying Position Effect Variegation in Drosophila Melanogaster

Bertrand Cinquin, Joyce Y. Kao and Mark L. Siegal

Contents

This Chapter has been reviewed by Jonas Øgaard, Research Institute of Internal Medicine, Oslo University Hospital.

What You Will Learn in This Chapter

Many of the methods developed for the analysis of bioimages focus on microscopy images on the cellular level. However, bioimages can also be used by biologists to assess non-cellular level morphological phenotypes. Collecting non-cellular images and developing image workflows for them is similar to working with microscopic images, but also has its unique challenges. We hope to impart upon the reader the following:[1]

1. Why images and workflows are necessary for improved assessment of subjective phenotypes (e.g. shades of color);
2. Which points to consider when collecting color images;
3. How to incorporate an *ilastik* segmentation model into an ImageJ macro;
4. One example workflow illustrating how to derive metrics for spatial patterns.

7.1 Introduction

Often times morphological phenotypes are subjectively scored manually, based on visual inspection, using a rating scale for either the severity of a characteristic or simply the presence or absence of a feature. However, in the cases where images exist, the pixel values and patterns of distribution can be more accurately measured using automated algorithms leading to more fine-scaled analyses, which was the original motivation for the work done in this chapter.

7.1.1 What Is the Big Deal with Color Images and Fly Eyes?

Color is a phenotype that has been used throughout the history of genetics research. The birth of modern genetics began with Mendel's systematic experiments with the colors of the flowers and seeds of the pea plant, but even before this, farmers were selecting on color for livestock and crops (Mendel, 1866). In this chapter we will focus on the popular model organism, *Drosophila melanogaster*, the fruit fly. One of the first phenotypes described in *D. melanogaster* involved a mutant of the sex-linked gene, *white*, which encodes for an important intermediate product that leads to the red pigmentation in fly eyes. Mutant flies have white eyes as opposed to wild-type red eyes (Morgan, 1910). Following the discovery of the *white* gene, a number of other eye-color mutants were also discovered (Morgan, 1911) (See ◘ Fig. 7.1).

More complicated eye-pigmentation mutants arose as more and more genetic tools were being developed in the fruit fly model. One mutant of note (and the focus of this paper) is the w^{m4} mutant (Muller, 1930). The w^{m4} mutant is a classical example of position-effect variegation (PEV). An inversion on the X chromosome relocates the *white* gene next to pericentric heterochromatin so that the neighboring chromatin state determines whether or not the *white* gene is expressed. When the neighboring chromatin is in the euchromatic state, *white* is expressed, whereas in the heterochromatic state, *white* is silenced. These alternate states are subject to random cellular events

---

1 This chapter was communicated by Jonas Øgaard, Research Institute of Internal Medicine, Oslo University Hospital, Norway.

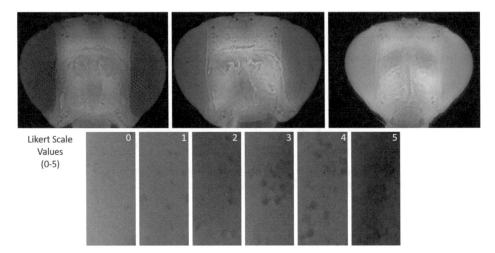

Fig. 7.1 Images of a fly head showing eyes of different colors (*top row*) and rectangular eye 'swatches' cropped from head images of w^{m4} mutants, which have mottled eye color. The darker 'spots' in these eye images are pigmented eye cells/patches and a corresponding Likert scale value of the degree of patchiness/pigmentation is indicated on each image. (*bottom row*)

during eye development, so that patches of cells with different chromatin states exist in the same eye. The result is eyes with mottled or variegated patterning as shown on Fig. 7.1, with some eye cells expressing *white* and therefore red pigmented, whereas other eye cells have silenced *white* and thus are white. Other PEV mutants include bw^{VDe2}, which places pericentric heterochromatin next to the *brown* (*bw*) gene, resulting in variegated brown pigment in the eyes (Sass and Henikoff, 1998).

PEV mutants have been used to indirectly assess overall changes in chromatin regulation in genetic experiments, with more pigmented eye cells implying *'looser'* chromatin, and fewer pigmented eye cells implying *'tighter'* packed chromatin.

7.1.2 How Is PEV Quantified Now and Potential Issues

Different methods to assess the amount of *"variegated-ness"* have been developed over time, starting with pigment extraction and quantifying the pigment using spectrophotometry (Ephrussi and Herold, 1944). However, the reliability of this approach has been questioned and the current method proposed by Sass and Henikoff (1998) uses an experimenter-defined ranking system (or Likert scale) for the extent of red pigment based on visual inspection. Additional safeguards for reproducibility were built into the method by establishing the five-rank Likert scale on independent samples of wild-type and mutant flies and adding a second experimenter to help define the scale and independently score the flies in the same order as the first experimenter. Scores from both experimenters are then averaged together for analysis.

The Likert scale (LS) approach has been successfully used in previous studies to quantify modifiers of PEV, and is generally a popular method to quantify the intensity of a visible phenotype. Although there is nothing inherently wrong with this approach, it might not be the most appropriate in all situations. Although the scale has an adaptable number of ranks (five, six, ten, or even all the way up to one million, etc.), the LS in reality has far fewer "effective" ranks. Users of a LS tend to agree in score at the extreme ends of the scale, but there is less agreement among the middle scores. Although losing "effective" ranks is not catastrophic when phenotype modifiers are one or two genes of large effect, it could prove to be problematic in a case in which phenotype differences must be detected with fine resolution. An example of such a case is when there are many modifiers of small effect and one hopes to identify these modifiers using a quantitative-trait locus (QTL) mapping approach. By the law of large numbers, more precise Likert scores can be achieved by more independent users rating each specimen, or by increasing the sample size of specimens. Both solutions require considerably more effort and time.

7.1.3 The Fallacy of Human Perception and Why Automated Analysis of Images Is King

Color is a deceptively easy phenotype to score. Human vision, especially perception of color, is itself a highly variable trait determined by the underlying combination of molecular/genetic mechanisms (Deeb, 2005) and neural processing of visual signals (Schlaffke et al., 2015). Defects in the former can result in different forms of color blindness, which can affect a person's ability to correctly judge a color phenotype. The latter however should be more worrying because people can often be intentionally or unintentionally tricked into incorrectly perceiving color by their own biology.

In the age of digital images, we can bypass the color-scoring biases produced by the subjectivity and variability of human perceptions and improve upon the labor-intensive gold standard of scoring PEV by visual inspection using a LS. Using images and automated analysis, we can obtain objective and consistent measurements faster via computer vision. We therefore lay out in this chapter an automated method that can quantify eye color from images of fly heads using a commercially available imaging setup and open-source analysis software. In addition, we explore ways to more precisely quantify position-effect variegation with additional spatial metrics to assess the "patchiness" of variegation. The spatial patterns of the patches might be biologically significant in determining how randomness enters the developmental process.

7.2 Dataset

For this chapter, the dataset includes 20 images taken with brightfield microscopy of the heads of male progeny produced by crossing w^{m4} *Drosophila melanogaster* females to either of two different PEV modifier mutant males. Some of the heads are heavily 'patchy' and others are not (i.e. variable variegation). ◨ Figure 7.1 is representative of what the image set looks like.

7.2.1 Imaging Conditions

For illuminating each fly head, as with many forms of imaging, a fixed lighting source should be used and care needs to be taken that it is the only light source in the environment. Natural light, as well as overhead room lighting, can unintentionally add noise and shadows in images. Depending on the source of fixed lighting, intensity of the light could be affected by factors such as lamp warm up time.

7.2.2 About Image Acquisition, Preprocessing, and Color Normalization

We used the Keyence VHX-1000 microscopy system for our image acquisition. All fly heads were imaged at $400\times$ magnification for a 0.6 s exposure. Before processing each batch of specimens, the camera was white balanced with a standard white card (Vello white balance card set). In addition, we acquired both a darkfield and a brightfield background image to perform corrections against camera sensor noise (e.g. "hot pixels"), as well as variable background illumination intensity (i.e. flat-field correction; Landini, 2020). Finally, we imaged an 18% grey card (Vello white balance card set) for color normalization across imaging batches.

Heads are not flat and the depth of field (focus) is shallow when imaging small objects up close, so we take image stacks of each head and apply a focus-stacking process to the stack to produce a single image that has a greater depth of field, and thus the entire head is in focus. On the Keyence VHX-1000 microscope, stack acquisition and focus-stacking are automated, resulting in a single fully-in-focus TIF image of each head. However, any microscope that can take image stacks can be used, and open-source image processing software packages such as ImageJ have components for focus-stacking.[2]

7.2.3 Dataset Download

Image data used in this chapter can be downloaded from our Zenodo repository (▶ http://doi.org/10.5281/zenodo.3975644) or from the GitHub repository associated with this book.

The ZIP package contains several folders. The first one called "Data" contains two folders relative to the two different investigated mutants. A second folder called "Processfiles" contains some files used in preprocessing for white balancing and dead pixel corrections, the ilastik training file and the full code. Note that due to limitations of the ilastik plugin it is important to save the ilastik-project file (.ilp) in a folder structure without spaces in the path.

Once the data is downloaded, we can do a quick pre-assignment before we dive into the workflow. Looking at the raw non-processed sample images, appreciate the variability of patchiness of the fly eye; as it is commonly said—"beauty is in the eye of the beholder." Use the subjective Likert scale approach to create a base set of 'manual scores' and rate the 'patchiness' of the left and right eye images on

---

2   i.e. Focus Stacker plug-in ▶ https://imagej.nih.gov/ij/plugins/stack-focuser.html.

a scale of 0 (no patches) to 5 (very patchy). After this chapter, we can compare these subjective evaluations against the objective measurements obtained from this chapter's automated processing and analysis macro.

7.3 Tools

- Fiji, Schindelin et al. (2012)
  - Download URL: ▶ https://imagej.net/Fiji/Downloads
- ilastik, Berg et al. (2019)
  - Download URL: ▶ https://www.ilastik.org/download.html
- ImageJ plugin ilastik last updated version (we used version 1.3.3)
  - Use update site function to install this plugin

7.3.1 ilastik Configuration in ImageJ

The ilastik plugin for ImageJ needs to be installed and ensured to be up-to-date. Before running the macro command, be sure that the plugin is properly configured. In the submenu of the ilastik plugin, select **"Configure ilastik Executable location"**. You will be asked to choose the file ilastik.exe, the number of threads to use (4, use −1 to use the maximum number), and amount of RAM to dedicate to the task (4 GB is more than enough). As we will work with (relatively small) 2D images, there is no need to allocate more than a few gigabytes of RAM.

7.4 Workflow Overview

The workflow overview (▣ Fig. 7.2) is described below. Fully automatic steps and steps that require user interactions are separately labeled. References to the lines in the full macro code found in the code repository of this book are also provided.

1. Workflow preparation
   - Asking user to input working directory *(User Interactive)* [line 6–19];
   - Listing all files in user directory, preparing empty arrays for data collection, and finally opening the image of the fly head *(Automated)* [line 23–30; 209–233];
2. Crop the left and the right eye areas *(User Interactive)* [line 50–67; 181–207];
3. Use ilastik to perform pixel-based segmentation of the fly eye and get binary masks *(Automated)* [line 70–74];
4. Use the binary masks to extract relevant information concerning the eye and its patchiness *(Automated)*:
   - Part A: Simple metrics [line 79–128]
     - Analyze Patch Area
     - Analyze Patch Intensity;
   - Part B: More advanced metrics

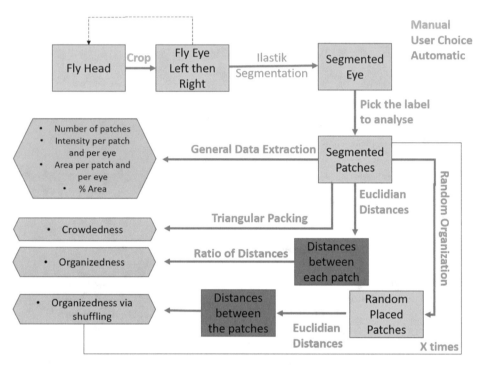

□ Fig. 7.2 Schematic of the workflow, with boxes colored in orange when the code is dealing majorly with images, in grey with arrays, and in green when it computes the features which are stored in the output tables and used in the subsequent analysis

  – Assess **'Crowdedness'** via calculating the Maximum Triangle Packing Value [line 137, 346–353]
  – Assess **'Organizedness'** using distances between patches [line 139–179, 268–344];
5. Export the calculated features [line 356–389];
6. Batch processing with multiple folders (application of Chapter 1) [line 23–30; 209–233].

The subsequent sections describe in more details each of the steps included in the workflow.

7.5 Step 1. Workflow Preparation

7.5.1 Selecting the Working Directory

We start with a very simple user interface that will ask for the directory containing images of the whole fly heads. We present two options: to use the getDirectory function, or the #@ File script parameter.

The getDirectory works in one step by calling the function that will directly open a path selection window. The #@ File option is more customizable and works in several steps. A user interface will open with prompt to select a working directory, which can be set using a *Browse* button. Once selected, the macro will continue with the selected directory by clicking *Ok*. By pressing *Cancel*, it will stop right away. Both methods are valid, but the former does not allow prompts for the user and this can get confusing if a workflow requires multiple sources of files and additional input (e.g. output folder, labels, etc.), which can be handled in one window with `#@ File`. Thus we find `#@ File` the more intuitive for the user since it enables to provide hints to the user for what to select. For more information on script parameters, please refer to the Batch Processing chapter (Ch. 1) in this book.

```
1  DirSrc = getDirectory ("Choose DataSource Directory");
2  //or
3  #@ File (label = "Source of Raw Images", style = "directory") DirSrc
```

We use the working directory to gather the list of files to analyze. For now, all the images are contained in one folder. At the end of this chapter, we will present a solution to navigate subfolders.

7.6 Step 2. Cropping Left and Right Eye Areas

Once an image is open, the next step is to implement a preprocessing cropping step to sidestep an issue of segmenting the eye from the *'face'* of the fly. Segmentation by thresholding is difficult in cases when the fly eye color is closer to *white* such that the boundary between the eye and the *face* is almost indistinguishable. Thus we simplify this step and crop rectangles out from the eye regions for further analysis. This is illustrated in ◘ Fig. 7.3. In our provided data set, our rectangular cropped image of an eye is 184×416 pixels in size because this was determined to be the largest rectangle that fits in the area of the fly eye in the images of the whole fly head. In general, the cropping could be of any (suitable) size or shape.

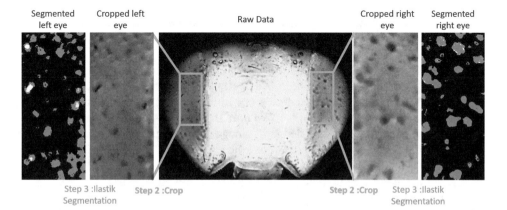

Segmented left eye   Cropped left eye   Raw Data   Cropped right eye   Segmented right eye

Step 3 :Ilastik Segmentation   Step 2 :Crop   Step 2 :Crop   Step 3 :Ilastik Segmentation

◘ **Fig. 7.3** Illustration of the Cropping (Step 2) and Segmentation (Step 3). Rectangular areas are cropped from the left and right eyes and ilastik is used to perform the segmentation (1. black = background, 2. white = debris, 3. grey = pigmented patches)

? Exercise 1: Write a Cropping Macro Using the Help and Hints Below

Write a small function that marks the region of interest (ROI) on the main image before cropping it.

- Display a rectangular ROI;
- Allow the user to reposition;
- Crop the rectangle.

Hint #1: To pause the macro and let the user do the ROI selection, one can use the command `waitForUser`. We use it as a direct interface between the code and the user, but it could also be used, in principle, as a hacky breakpoint to pause macros for debugging code.

Hint #2: In Fiji, we *crop* by duplicating ROIs. Thoughtful naming of the duplicated image can help organize and describe what is in the image. We propose to attune the original file name to reflect which eye has been cropped. Therefore the function needs to handle another argument.

Extra credit: Save the duplicated image as a file in the working directory. We can entertain the possibility that the function could check if the cropped image of the eye exists in the working directory. If it does already exist, the function can ask the user if they wish to crop again in case the crop was not satisfactory in the previous attempt. The solution to this exercise can be found at the end of the chapter.

7.7 Step 3. Segmentation by Using Ilastik

With the cropped images now saved, we can move forward with segmentation by using ilastik. We use here the simple Pixel-based Segmentation option in the ilastik software (Berg et al., 2019), which can be run with the following command:

```
run("Run Pixel Classification Prediction", "projectfilename=" +
↪   DirCorr + "\\" + TrainingFileName + "inputimage=Left
↪   pixelclassificationtype=Segmentation");
```

It will take a few seconds to get the segmented image, but afterwards is where things get interesting; we can now start extracting information from the images. An example of segmentation is shown in ◻ Fig. 7.3.

We are not going to go into details regarding using ilastik itself here, because great tutorials are available on the ilastik website.[3] We do not provide a training set for the performed segmentation, but instead provide a pre-trained ilastik model as part of the accompanying materials to this chapter. We note that training is laborious and requires some *Drosophila* domain knowledge in recognizing features in the images. However, we will make some general points about the training process itself and how segmentation works in ilastik.

ilastik segmentation is based on machine learning and requires selecting features, and training a model to categorize pixels into different segmentation classes. Training ilastik models requires an annotated training set of images, which can be created by the user within the software. In short, the user assigns labels to pixels using a very intuitive interface where they simply draw or 'color' on a training image. We

---

3   ▸ https://www.ilastik.org/.

found that the accuracy of the annotation was dependent on how familiar the user was with the subject matter in the images, with more experience leading to more accurate annotation. To minimize the effect of variations in the subsequent analysis due to different levels of the domain knowledge (affecting quality of annotations), we provide a trained model as a part of this workflow. In our case, we have three output classes: the background (i.e. white), pigmented patches (i.e. red), and debris (e.g. bristles and flecks that were not successfully cleaned off pre-imaging).

Images segmented using the provided ilastik model will have new pixel values corresponding to their label numbers. The pixels with the value 0 are pixels classified with label 1 (background), those with the value 1 are classified with label 2 (debris), and those with the value 2 are classified with label 3 (pigmented patches). Selecting a specific pixel value (i.e. specific label) is easy with the thresholding tool in ImageJ with the lower and upper bound set to the desired pixel/label value.

```
setThreshold(LabelNumber, LabelNumber);
```

Being able to select labels now allows us to extract information on relevant features in the image. We will see how to retrieve intensity information from the original images in a bit.

7.8 Step 4. Extracting Measurements from the Segmented Objects

7.8.1 Part A: Simple Metrics Using [Analyze Particles...]

Running [Analyze Particles...] will give all the information we are going to need for the next parts of the analysis workflow. We will set up the measurements by selecting which metrics we want to extract. For our workflow, the *intensity "Mean Gray Value"*, the *Area*, and the *X and Y coordinates ("Centroid")* of each patch are important. When running [Analyze Particles...], we can also select *Summarize* to display additional *Results* tables including the number of particles (pigmented patches, in our case with fly eyes), the total area covered by the particles, and the percentage of pigmented pixels. [Analyze Particles...] will fill the roiManager. When we use [Analyze Particles...] on the segmented image (or 'binary' masks), it should be noted that the intensity of the segmented particles is not a true reflection of the intensity of 'color' in raw image because the value of pixels in the segmented image (resulting from ilastik) corresponds to the output class label number we defined with image annotation and model training. Therefore, to retrieve the original intensity of each ROI, another measurement must be performed on the raw cropped image, however using the ROIs obtained from the segmented image (the masks obtained from ilastik).

As with many tasks in image analysis, these metrics can also be retrieved in other ways. For example, to get the number of patches, the command roiManager("count") will also work. Additionally, the total area can be calculated by looping through each ROI, calculating the area, and updating a variable that keeps track of the total area. Dividing the total area by the number of ROIs gives the average size of each ROI. The percentage of pigmented pixels is the total area of the ROIs divided by the total size of a labelled image and multipled by 100 (i.e. the size of the labelled image is defined as the size of the cropped rectangle—ROI—which is 184*416 pixels in our particular case).

The code below shows the essence of the use of roiManager to perform measurements iteratively over differently labeled objects (different pigments), and storing measured values in arrays.

```
1   Areadata = newArray(roiManager("count"));
2   Meandata = newArray(roiManager("count"));
3   Table.rename("Measurements","Results");
4   selectWindow(Side);
5   TotalPatchArea = 0; TotalIntensity = 0;
6           for (j=0; j<roiManager("count"); j++) {
7                   roiManager("select", j);
8                   roiManager("measure");
9                   Meandata[j] =
    ↪    getResult("Mean",j+roiManager("count"));
10                  Areadata[j] =
    ↪    getResult("Area",j+roiManager("count"));
11                      TotalPatchArea = TotalPatchArea + Areadata[j];
12                      TotalIntensity = TotalIntensity + Meandata[j];
13          }
```

GeneralMeasurements.ijm

In addition to what was previously mentioned, we decided to build two arrays with length equal to the total number of patches. We will store the area of each patch in the first one, and the intensity of each patch in the second one. We will discuss the results tables later.

7.8.2 Part B: Crowdedness and Organizedness

Information about the number, average size, and percentage area of patches does not necessarily reveal much about how these patches are distributed in our eye images. We illustrate in ◻ Fig. 7.4 a case where these metrics are not enough. Let us take the example of eye images A and B, with 10 pigmented patches each, and equal total eye image area. The patches are all equal in size, but in eye A the 10 patches are arranged in a line, whereas in eye B the patches are randomly distributed across the eye. The number (10 patches), average size (100 pixels), and percentage of area covered by patches (2%) are exactly the same between eyes A and B and thus we cannot capture the difference between the distributions of the patches—the straight line versus random pattern in eye A and B, respectively, using the metrics introduced so far in this chapter. Hence, we need other metrics to describe the spatial distribution of the patches and capture this difference in patch arrangements.

Understanding spatial patterns of the patches can be biologically significant in assessing the randomness of the developmental mechanism and help us pose stronger hypotheses regarding its underlying nature. Many spatial distribution analysis methods exist already; in this chapter, we walk through how to derive some simple metrics:

— *Crowdedness*: This metric will essentially take into account how large the pigmented patches are and how many more we could fit into our total image. This

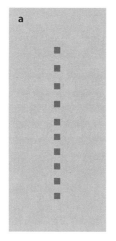

Number of patches = 10
Average size of patches = 400 px
% of pigmented pixels = 2

Number of patches = 10
Average size of patches = 400 px
% of pigmented pixels = 2

☐ **Fig. 7.4** Hypothetical examples of patches where basic metrics are exactly the same, but appearance is vastly different

one metric roughly describes what a combination of number, average size, and percentage of pigmented area describes, with an advantage that it is much easier to comprehend one metric than an array of 3 metrics when performing statistical analyses.

– *Organizedness*: This metric evaluates Euclidean distances between (the centroids of) the pigmented patches (ROIs) to assess how they are organized in the 2-D space of our image. We can compare what we calculate from our images with theoretical organizedness values from literature to assess how far from an ideal organization the patches are in 2-D space. We will also present a second way of assessment using a statistical method called a permutation test to generate a distribution of simulated organizedness metrics, which we can then compare our actual computed organizedness metric with, to get statistical significance (e.g. a p-value) and assess how far the considered organization is from a totally random one.

7.9 Deriving Crowdedness

7.9.1 First compute the Max Triangular Packing Value

The *Max Triangle Packing* value (MTPV) gives how many of the patches can fit in the studied area. We have already computed the percentage occupancy for the patches in the area in Part A. This number is a proxy of how 'crowded' the eye is with patches. However, we are not taking into account that the eye is a compound eye composed of individual units called ommatidia, and the distribution of these patches can be influenced by these repeating units and the developmental process that produces them. By solely taking the percentage, we do not capture the information of how these patches are spatially arranged. To take the repeated eye units into account, we assume pigmented patches to have a perfectly circular shape, and the same area (and so the same radius), and we "pack" these circles into the cropped eye image, as illustrated in ◘ Fig. 7.5.

This is not a perfect estimate because the eye is three-dimensional. Even though all eye units are equal in size, units viewed from different angles due to the curvature of the eye will be different-sized pixel-wise in the 2-D images. Additionally, we do see in the bottom row of images in ◘ Fig. 7.1 that one eye unit can contain several patches. However, we make an additional assumption here that the view plane is flat and one patch equals one eye unit in our first triangle packing approximation. These assumptions can be improved upon in future versions.

The simplified pattern of organizing the maximum number of circles in a rectangle, which is considered the 'optimal' arrangement is called *triangular packing*. The total number of circles that fit is our *Max Triangle Packing* value—MTPV. Let us now work through how to come to the 'optimal' arrangement, in the following exercise.

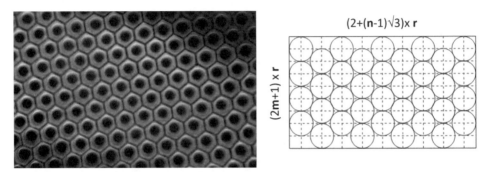

◘ **Fig. 7.5**   Triangular packing: what *Drosophilla* eye cells would look like if flattened out in 2D space (left), and the visual depiction of how we pack perfectly round patches into a rectangular space—length and width are expressed in terms of the number of rows *m* and columns *n*, and the average radius *r* of a hypothetically circular patch

? Exercise 2: Write the `TrianglePacking` Function for Our Rectangular Cropped Eye Image That Is 184 × 416 Pixels in Size

Hint #1: First, determine, for a given average patch size, what is its radius, r, if it were a perfect circle. Here is the hint: Work backwards using the formula for the area of a circle, πr^2, from the average area of a patch that was computed earlier in this chapter (Step 4; Part A). It is the *TotalPatchArea* divided by the number of patches.

Hint #2: With this radius, r, how many circular patches can fit in our region of interest? Break it down to how many circles can fit across both dimensions (width and length-wise). Use the diameter.

Calculating Crowdedness

To calculate 'Crowdedness' we need to determine the ratio of the number of patches we have segmented (and previously counted with [`Analyze Particles...`] and MTPV:

$$Crowdedness = \frac{Number\ of\ patches}{Max\ Triangle\ Packing\ Value} \tag{7.1}$$

7.10 Assessing Organizedness

For this particular metric which we will now dive into, we borrow some theoretical basis of organizedness from (Audet et al., 2010); it discusses the mathematical optimality of points arranged in 2D space.

7.10.1 Computing Pairwise Distances

We want to measure the pairwise distance between the patches. For that purpose, we will compute the Euclidean distance between the centroids of the patches. We got the coordinates of the centroid of each patch from [`Analyze Particles...`].

In the next set of exercises with hints following each prompt, we walk through how to compute the pairwise distances. Solutions can be found at the end of the chapter.

? Exercise 3.1 : Write the Steps to Iterate Through All Pairwise Patches

We can use a drawer of n socks to illustrate how we perform counting of pairs. How many different pairs of socks can be made from the drawer? (Note: Any pair should be counted, there are no "matching" and "non-matching" socks, and no difference between "sock 1" and "sock 2".) To count these in a structured way, we take out one sock and count all pairs we can make of it and the $n-1$ remaining socks in the drawer. Then, we set the first sock aside and pull out a second sock and count the number of pairs we can make with it and the remaining $n-2$ socks left in the drawer. We move on to the third, fourth, fifth socks in the same manner, pairing it with the decreasing $n-3$, $n-4$, $n-5$ socks left in the drawer. This pattern leads to the number of possible pairings, as the sum of $(n-1) + (n-2) + (n-3) +...+ 1$, which simplifies down to $n*(n-1)/2$ total combinations or, in combinatorics notation, there are $\binom{n}{2}$ ways to select two elements

out of *n* elements. Use the *sock drawer algorithm* to write down the steps needed to make pairwise combinations of the patches.
Hint: We need two `for` loops when we are considering pairs of objects. Also, be mindful of the indices!

❓ Exercise 3.2 : Build in a Way to Keep Track of Distance Calculations

Since we know the total number of combinations, the calculated pairwise distances will be put in an array of a size of $n*(n-1)/2$. To fill the array in a progressive way, we can use a counter or index variable that will increment by one each time a distance is calculated. We would also like to keep track of which two patches were used to calculate each distance. Now write down the steps to achieve this.

❓ Exercise 3.3: Write the Code to Calculate Pairwise Distances

Now translate the step-by-step instructions from the *sock drawer algorithm* into code in the form of a function named `DistanceAnalysis`.

7.10.2 Ratio of Maximum and Minimum Distances (*r*)

For a given number of patches, the way that the patches are organized can be defined as mathematically 'optimal' if the distance between each pair of patches is maximized as detailed in Audet et al. (2010). Adding to this basis, the 'most optimal' or most organized arrangement is when the ratio of the maximum distance to the minimum distance between patches reaches a minimum. It is perhaps easier to comprehend this by looking at ◧ Fig. 7.6, which gives a few examples of such 'optimal' organizations with red lines being maximal distances and blue being minimal distances. It should be noted that there is also an additional constraint in our use-case with regards to the arrangement of the patches, since the arrangement must fit within the space and shape of the cropped eye image.

❓ Exercise 4.1: Compute a Summary of Organizedness, *r*

We already have all the distances between all the patches that have been computed. To get a summary min-max ratio, and by that a sense of overall 'degree of organization' of all the patches, we will use the average maximum and average minimum distances considering each pair of patches. We will build arrays that contain the maximum distances for each found patch and the same for minimum distances, take the average of each array and then compute the general min-max ratio. As we previously built a nice table containing the distances and the indices for each point, it should not be too difficult to implement this.
Hint 1: To write the structure to go through all the distances, we need to use a `for` loop to go through all the patches, and then a second `for` loop to go through all the computed distances, with the right `if` statement.
Hint 2: To take the mean of an array, we can sum all the terms and divide by the number of terms. Alternatively, we can use the command/function `Array.getStatistics`.

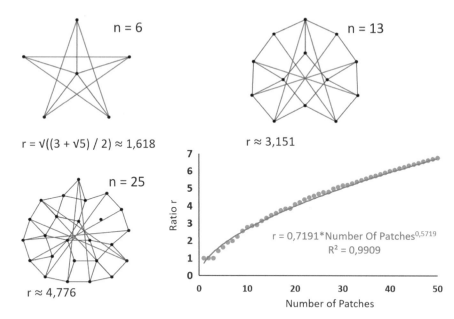

Fig. 7.6 Examples of optimal arrangements of n points, with indicated maximal distances (red) and minimal distances (blue) in different configurations. Their ratios of Maximum and Minimum Distance (r) and a graph presenting the relationship of the ratio r and the number of patches are also shown. A fit by a power function should be used when there are more than 30 patches, which is shown in the plot (bottom right)

Exercise 4.2 : Compare the Min-Max Ratio with Theoretical r Values to Assess Organizedness

Now that we have a summary min-max ratio, we should use it to assess the degree of organization. We compare our min-max ratio to theoretically ideal ratio values provided by Audet et al. (2010) by looking up the ratio that corresponds to the number of patches we have. For the purpose of our exercise, we hard-coded the theoretical 'optimal' ratios from Audet et al. (2010) in an array contained in the file, *Min_MaxMinDistRatio.csv*. We can then simply look for the right index corresponding to the number of patches and make a comparison via computing a ratio of the observed versus actual values. A value close to 1 would indicate a high degree of organization. That being said, the ideal ratio has been derived for only a small number of points. For any number above 30 points (or patches in our case), we compute an approximation using a power law function as it gives the best fit ($R^2 = 0.99$) to link the expected minimal ratio r between the maximum and the minimum distance for a set of points spread in a two dimensional space and the number of points (or patches) of this set. The resulting approximation is $r = 0,7191 * [NumberOfPatches]^{0.5719}$, which can be seen in plot form in ■ Fig. 7.6.

Comparison with a Distribution of Randomly Shuffled Patches

There are alternative ways to assess organizedness in addition to looking at theoretical r values. One method we will go through now is to repeatedly shuffle the patches and measure the difference in the Minimal ratio of maximum and minimum distances, for these different patch distributions. We are essentially checking if the distribution we have and the corresponding computed min-max ratio is an outlier value, compared to the distributions of min-max ratios of random arrangements in our cropped eye image. If the patches in an eye are in a clustered pattern, that ratio will be significantly different from the ratios of random patch arrangements. In statistics, the method we are using here is called a permutation test. Implementing a random shuffling of objects seems fun and coding it could be an interesting challenge, if some guidance is provided.

Fiji has a command `moveROI` that takes a desired displacement as an argument. Giving a new set of coordinates for the patches can easily be done by using the command `random` that returns a number between 0 and 1. Multiplying this random number by the height of our cropped eye image and then repeating the same procedure with another random number and the width, gives random displacement to relocate our patches. However, the new set of coordinates needs to fulfill some constraints for the ROI to be valid.

? Exercise 5.1: Come up with the Two Conditions That Must Be Checked When We Move Around ROIs

Hint 1: Think about what happens when the centroid of a ROI which is 5 pixels wide gets moved to coordinate (0, 0).

Hint 2: Now move another ROI to (0, 1), next to the assumed 5 pixel-wide ROI at (0, 0), as mentioned in Hint 1. What do we observe?

? Exercise 5.2 : Write the Function `random()` to Check the Conditions Identified in the Previous Exercise and Then to Randomly Move Around the ROIs

Hint 1: We can use a `do...while` loop to implement this randomization, where the code is executed as long as the while condition is true. We can write it in such a way that the loop keeps running (i.e. a new set of coordinates is generated) if the new set of coordinates does not satisfy the conditions defined in the previous exercise. This loop needs to be done for each ROI patch.

Hint 2: There is always a risk that, for a given arrangement of patches generated by some moment during the randomization, there may be no way to place the next patch without violating the overlapping and out-of-bounds conditions defined above, and we are stuck in an infinite loop. To remedy this, the function should be aborted if too many attempts are made. We add a `testcount` variable in the `do...while` loop to force an exit when it looks like there will be no solution, based on the number of placement attempts. At this point, we will need to restart the randomization process by creating a new empty image and starting over. The hope is that we will not repeat the same arrangement that led us to the infinite loop in the first place.

? Exercise 5.3 : Write Down the Steps to Generate a Distribution of Summary Min-Max Ratios

We use the new, randomized set of coordinates of the patches and compute the pairwise distances as well as the degree of organizedness with our previously coded functions, `DistanceAnalysis()` and `MaxMinRatio()`.

Once we have a randomized distribution of min-max ratios, we can compute a p-value here by counting the number of r values greater than the actual summary r and dividing it by the total number of random r values generated overall. When we talk about p-values, we enter into the realm of hypothesis testing in statistics, which means we are making a call on whether or not we reject a null hypothesis. The null hypothesis in our case is that the min-max ratio we calculate from the patches in the cropped eye image is not in an *extreme* arrangement, meaning particularly large or small ratio. By generating random arrangements of patches like we just did, we are generating a distribution of min-max ratios where the arrangement is mostly 'random', whereas we get *extreme* arrangements when clustering or 'organized' patterns of patches occur, which result in extreme min-max ratios. Therefore, if our p-value is small (e.g. $p < 0.05$), we reject the null hypothesis, i.e., the hypothesis that the patch arrangement in our cropped eye image is in some sort of *extreme* arrangement.

7.11 Step 5. Exporting the Calculated Metrics into Tables

We create three different tables containing:
- the different calculated metrics for the overall cropped eye image (e.g. number of patches, average patch size, crowdedness, etc.;
- the area of each patch for each analyzed eye in such a way that distribution of areas can be retrieved;
- the intensity of each patch for each analyzed eye.

There are two ways of making a table. We either add a new row each time an image is analyzed or store the different arrays in such a way that they can all be put in a table as columns at the end, when all the images are processed. We use the former because we work with one eye image at a time and the speed of the analysis is not our main concern here. We will create different functions to fill up the tables. There is nothing difficult there besides keeping track of the indices. These steps are addressed in the main code using the functions `fillFinalTable` and `fillAreaTable`.

7.12 Step 6. Batch Processing and Further Considerations

Now that we have all the bricks, building the full macro should not be too much of an issue, besides a few additional thoughts. To summarize the whole workflow we presented in this chapter, please revisit ▣ Fig. 7.2.

 Exercise 6: Write the Step-by-Step Instructions for Batch Processing, Considering the Comments Collected in the Hint Below

Hint: We present a couple considerations to be kept on mind when stringing together the functions in batch processing.

7.12.1 A Fly Head Has Two Eyes

A very obvious first point: a fly has two eyes and both matter for the analysis. We have one input, the entire fly head image, which will turn into two when we crop the eye regions of interest. We want to ensure that the indices are consistent when filling the final table with the analysis results. In the table, the image of the head of the first fly with the index 0 (in the list of head images) will give two lines in the final table, one for the left and one for the right eye (i.e. rows 0 and 1). The second head image with the index 1 will give two new rows of results (i.e. rows 2 and 3) and so on.

7.12.2 Eyes That Have Zero or One Patch

When we come up with a general solution for a problem, we should also give thought to special cases that could potentially lead to errors, and handle those cases separately. For example, sometimes we encounter eye images that do not have patches or have only one patch. This would present a problem in computing pairwise distances because there cannot be a pair of patches. It therefore makes sense that, when only one patch is found or there are no patches, we do not calculate pairwise distances, or anything downstream that depends on the distances in the workflow.

Keep in mind we would need to add to our tables with results a *NaN* or *N/A* for these special cases when a measurement cannot be computed. Otherwise, we will encounter a mismatch of indices between the image being analysed and where the calculated metrics need to go in the tables.

7.12.3 Step 6.3 Batch Processing into Multiple Folders

Batch processing is rather simple once we build an array containing the paths for the files to analyse. The command `getFileList(path)` is doing exactly that and therefore a simple *for*-loop going through the array of files will work fine:

```
List = getFileList(MainDirectory);
openfile(List);
function openfile(TheList){
    for (i = 0; i < TheList.length; i++) {
        open(TheList[i]);
    }
}
```

The idea here is also that the folder ONLY contains images and no other files, otherwise errors will occur if ImageJ encounters a file type it cannot process. It can be more challenging to analyse files found in multiple folders. Therefore, it should be carefully thought out how to store images when generating them so that we do not need to spend time arranging and rearranging files for running image processing/-analysis workflows. Additional conditions can be added to check if the file has the right extension, size, etc.

```
23   function listFiles(dir,finalList) {
24     list = getFileList(dir);
25       for (i=0; i<list.length; i++) {
26         if (File.isDirectory(dir+list[i])){
27             listFiles(""+dir+list[i],finalList);
28         }
29         else {
30             finalList[NmbFile]= dir+"\\"+list[i];
31         print((NmbFile++) + ": " + dir+list[i]);
32         }
33     }
34   }
```

ListFile.ijm

As we want to build an array with all the paths for the files to analyze, we need to know how many files there are to initialize properly the array. We simply use a function with the same structure, but only to count the number of files to analyze as shown in the code above. Finally we can link everything with the right initialization of the different variables.

7.13 Visualizing Results: Presentation and Discussion

At the end of the workflow, we will get three tables:
- Area Distribution.csv contains the area of each patch for each analysed eye;
- Intensity Distribution.csv contains the average intensity of each patch for each analysed eye;
- FinalTable.csv contains the different metrics collected along the way (Number of Patches, Average Size, Average Intensity of the whole cropped eye image, Percentage of Area, Crowdedness, Ideal Ratio of distances, Deviation from Ideal, Deviation from Random Distribution)

The tables are a convenient output from Fiji. However, a graph can speak a thousands words (or numbers). We suggest using plotting packages in R, such as *ggplot2*, or in Python, such as the *seaborn* library or if one is extra ambitious and nitpicky, the *matplotlib* library. Plotting in Python is covered in Chapter 2 of this book. Graphs are a quick way to see the outcome of the macro we just coded. We can visually compare the metrics such as the number of patches, the average size, the 'crowdedness,' the average intensity or even the degree of organization (i.e. deviation from ideal ratio of the maximum and minimum distances or the deviation of the measured ratio from our randomized distribution). The distribution of the average intensity per patch shows differences between the two mutants.

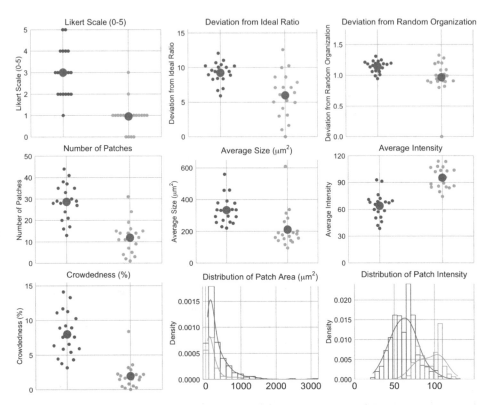

▫ Fig. 7.7 Strip plots illustrating the similarity in patterns between manually curated Likert scores (top left) and various metrics generated from coded workflow. Mutant 1 is represented in blue, and Mutant 2 in orange. The red dots are the average of the population

In addition to running the here developed workflow, the authors also scored, by visual inspection, the cropped eye regions using the Likert scale approach, to compare to the metrics generated by the workflow. ▫ Figure 7.7 illustrates our metrics generated by our coded workflow. The plots reflect the same patterns (or inverse pattern in patch intensity) as our manual scoring (top left strip plot). Having a coded workflow enables to tweak different parts of our workflow and rerun the code quickly and as many times as we want. This is not reasonable to do manually when the number of images to analyse is large. These plots in ▫ Fig. 7.7 validate what our eyes are able to perceive. We successfully translated our subjective perceptions into objective quantities!

Take-Home Message

We were able to extract and utilize relevant information from color (brightfield microscopy) images of fly eyes, to produce quantitative metrics for position-effect variegation.

To do so, we learned to use the existing functions and create our own metrics to adapt to our very own problem. We learned useful tricks such as how to pass the content of a variable from the function to the main body of the code.

This workflow can be adapted for problems beyond fly eyes. For example, understanding the spatial organization of objects in a given space is widely applicable in situations that include the foci in nuclei, lipid droplets in cells, endosomes in the cytoplasm, bacteria in an enclosed environment, etc.

The overall goal of this work was to translate our subjective qualitative observations (i.e. how patchy is this fly eye) into reproducible, quantifiable, fine-scale metrics (e.g. number of patches, intensity, organization,...) so that we can be more objective in measurements. This allows us to run more nuanced quantitative analysis to derive more precise conclusions in future studies.

Solutions to the Exercises

Each solution is included in the respective subfolder in the code repository associated with this chapter and book. The macros are intended for educational purposes and have been slightly tweaked from the full macro code, to work as stand alone exercises.

✅ Exercise 1: Write a Cropping Macro

The First Cropping Function

```
18   function crop_V1() {
19     makeRectangle(1254, 352, 184, 416);
20     waitForUser("Pause","Adjust the rectangle");
21     run("Duplicate...", "duplicate");
22   }
```

FunctionsCrop.ijm

The Second Cropping Function

```
25   function crop_V2(DirOut, FileName, Side) {
26     fss = File.separator;
27     if (Side == "Right") {
28       makeRectangle(1254, 352, 184, 416);
29       NewFileName=substring(FileName, 0,
         ↪   lengthOf(FileName)-4)+"R.tif";
30     }
```

Repeat for the other side

```
35     run("Duplicate...", "duplicate");
36     save(DirOut+NewFileName);
37   }
```

FunctionsCrop.ijm

The More Advanced Cropping Function

The first lines of this version are exactly the same as the second version; the difference appears in the very last lines, where we take the decision to crop, recrop or not recrop.

```
50  if(File.exists(DirOut+NewFileName)==1){
51    Answer = getBoolean("Cropping for this eye has been done, \n
        ↪  Would like to redo it ?", "Yes", "No");print(Answer);
52    if (Answer ==1){
53      waitForUser("Pause","Adjust and Update "+Side+" eye ROI");
54      run("Duplicate...", "duplicate");
55      save(DirOut+NewFileName);
56    }
57    if (Answer ==0){
58      open(DirOut+NewFileName);
59    }
60  }
```

FunctionsCrop.ijm

✅ Exercise 2 : Write the `TrianglePacking` Function

```
10  function TriangularPacking(PatchNumber, AverageArea) {
11      AverageRad = sqrt((AverageArea)/(PI));
12      print("The average radius is ",AverageRad);
13      NumberLines = floor((416/AverageRad-1)/2);
14      NumberColumns = floor((((184/AverageRad-2)/sqrt(3))+1));
15      NumberMax =
        ↪  (NumberLines-1)*NumberColumns+NumberLines*NumberColumns;
16      Crowdedness = PatchNumber/NumberMax * 100;
17      print("For a patch with an average area of 335.455 pxlÂ² in an
        ↪  area of 416*184 pxlsÂ²,\n we can expect to put
        ↪  "+NumberMax+ " patches. \n Therefore, the crowdedness is
        ↪  "+Crowdedness +" %");
18  }
```

FunctionTriangularPacking.ijm

✅ Exercise 3.1 : Write the Steps to Iterate Through All Pairwise Patches

The first loop needs to run *n-1* times and the second loop needs to run one iteration less than the first loop and so on, until considering the final distance between the object *n-1* and the object *n*. The second loop needs to start at the index of the first loop.

```
1  For (i=0; i< PatchNumber-1, i++){
2      For (j = i+1; j< PatchNumber, j++){
3          //Measure the distance
4      }
5  }
```

✅ Exercise 3.2: Build in a Way to Keep Track of Distance Calculations

```
1  //Fill up the X and Y arrays.
2  //Initialise the counter/the arrays/the tables
3  For (i=0; i< PatchNumber-1, i++){
4      For (j = i+1; j< PatchNumber, j++){
5          //Measure the distance
```

```
6        //Put the number into the array at the appropriate index
         ↪   (counter)
7        //Fill up the tables with the indexes i , j and distance
8        //Increment the counter by 1
9      }
10   }
```

✅ Exercise 3.3 : Write the Code to Calculate Pairwise Distances

```
25   function DistanceAnalysis(PatchNumber,X,Y, TableName){
26     Table.rename(TableName, "Results");
27     selectWindow("Results");
28     DistancesNumber = PatchNumber*(PatchNumber-1)/2;
29     Distances = newArray(DistancesNumber);
30     DistancesSorted = newArray(DistancesNumber);
31     counter2 = 0;
32     for (i = 0; i < PatchNumber-1; i++) {
33       for (j = i+1; j < PatchNumber; j++) {
34       Distances[counter2] = sqrt(pow(X[i]-X[j],2)+pow(Y[i]-Y[j],2));
35       setResult("Patch1 NÂ°", counter2, i);
36       setResult("Patch2 NÂ°", counter2, j);
37       setResult("Distances", counter2, Distances[counter2]);
38       counter2 = counter2+1;
39       }
40     }
41     Table.rename("Results",TableName);
42   }
```

FunctionDistances.ijm

✅ Exercise 4.1 : Compute a Summary of Organizedness

```
1    //Initialise a counter and the arrays holding the distances
     ↪   regarding the patch i and the arrays holding the maximmum and
     ↪   minimum distances
2    For (j=0; i< PatchNumber-1, i++){
3        For (i = 0; j< DistanceNumber; i++){
4            if (getResult("Patch1 N°", i) == j || getResult("Patch2
             ↪   N°", i) == j){
5            DistObjecti[counter3] = getResult("Distances", i);
6            counter3 = counter3 +1;
7        }
8        }
9    //  Fill up the arrays
10   //  Reinitialise the counter to 0
11   }
```

✅ Exercise 4.2: Compare the Min-Max Ratio with Theoretical r Values to Assess Organizedness

```
12   function CompareRatio(PatchNumber,Ratio){
13     fss = File.separator;
14     if (PatchNumber>=50){
15         Min_Ratio = 0.6972*pow(PatchNumber,0.5845);
16       else{
17         Min_Ratio = getResult("Value",PatchNumber);
18       }
```

```
19        close("Min_MaxMinDistRatio.csv");
20      IdealRatio = sqrt(Min_Ratio);
21    print("Ideal ratio for "+PatchNumber+" patches is "+
      ↪   IdealRatio);
22    print("The computed ratio is "+Ratio);
23    print("The deviation from the theoritical value is
      ↪   ",Ratio/IdealRatio);
24    }
```

FunctionCompareRatio.ijm

✅ Exercise 5.1 : Come up with the Two Conditions That Must Be Checked When We Move Around ROIs

 — The new ROI will fit in the window
   – If the ROI goes over the boundary of the image, its area will be less than the actual ROI area.
 — The new ROI will not overlap with any other objects.
   – If the ROI overlaps with another region that has been validated and filled with a maximum value (255 in a 8-bit image), the average intensity of the this new ROI will be less than 255, it will be easy to conclude that it is not at a valid position.

✅ Exercise 5.2 : Write the Function `random()` to Check the Conditions Identified in the Previous Exercise and Then to Randomly Move Around the ROIs

```
1   //Create NewImage to hold the new ROIs and initialise the counter
2   for (i = 0; i < NumberofPatches; i++) {
3         do {
4             //Create random new centroids
5             //Move the ROI to new coordinates
6             //Add the ROI and measure Mean Intensity and Areas
7             //Delete the new ROI
8             //Increment counter
9         } while (AreaNewROI < Area[i]-1 || IntNewROI < 255 ||
          ↪   testcount >200)
10  //Move the new ROI int he new image and fill the arrays for
    ↪   further analysis of the new organisation
11  }
12  //Close the new image
```

The working code FunctionRandomize.ijm can be found in the subfolder 5_Shuffle.

✅ Exercise 5.3 : Write Down the Steps to Generate a Distribution of Summary Min-Max Ratios

```
1   //Initialisation of variables : Iteration number, an array to hold
    ↪   the computed ratios and a table for the new computed distances
2   for (i=1, i<= Number of iterations, i++){
3       function Randomize the patches //FunctionRandomize.ijm
4       function Measure the distances //FunctionDistances.ijm
5       function Compute the ratio of the maximum and minimum
        ↪   distances //FunctionMaxMinRatio.ijm
6       function Compare the ratios and the ideal ratio
        ↪   //FunctionCompareRatio.ijm
```

```
7        }
8    //Close
```

The actual code can be found in the full macro between the lines 147 and 164.

✅ Exercise 6: Write the Step-by-Step Instructions for Batch Processing

```
1    for (i = 0; i< number of files to analyse; i++){
2        for (j =0; j<2; j++){
3            if (j==0){Variable = "Left" ; }
4            if (j==1){Variable = "Right"; }
5        Call Function with index 2*i+j
6        }
7    }
```

CodeStructureLeftRight.ijm

Acknowledgements   We would like to thank the NEUBIAS community for their input on the workflow over the years as well as the lab of Ernst Hafen at the Institute of Molecular Systems Biology and ScopeM of ETH Zürich for their advice and instruction on imaging methods and technologies. This work was made possible in part by NIH grants 1F32GM116321-01A1 and R35GM118170.

Further Readings   If packing, organising objects inside a space in 2D or in 3D are of interest, we recommend the first chapters of a book dealing with perception and spatial organization: L.C. Robertson, *Space, Objects, Minds and Brains*. To advance to subjects with more metrics, Andrey et al. (2010) is a must read. An ImageJ plugin for spatial statistics already exists (▶ https://imagejdocu.tudor.lu/plugin/analysis/spatial_statistics_2d_3d/start.) and can be easily merged into our macro.

References

Andrey P, Kiêu K, Kress C, Lehmann G, Tirichine L, Liu Z, Biot E, Adenot PG, Hue-Beauvais C, Houba-Hérin N, et al (2010) Statistical analysis of 3d images detects regular spatial distributions of centromeres and chromocenters in animal and plant nuclei. PLoS Comput Biol 6(7):e1000853–e1000853. https://doi.org/10.1371/journal.pcbi.1000853, 20628576[pmid]

Audet C, Fournier X, Hansen P, Messine F (2010) A note on diameters of point sets. Optim Lett 4(4):585–595. https://doi.org/10.1007/s11590-010-0185-y

Berg S, Kutra D, Kroeger T, Straehle CN, Kausler BX, Haubold C, Schiegg M, Ales J, Beier T, Rudy M, et al (2019) ilastik: interactive machine learning for (bio)image analysis. Nat Methods https://doi.org/10.1038/s41592-019-0582-9

Deeb S (2005) The molecular basis of variation in human color vision. Clin Genet 67:369–77. https://doi.org/10.1111/j.1399-0004.2004.00343.x

Ephrussi B, Herold J (1944) Studies of eye pigments of drosophila. I. Methods of extraction and quantitative estimation of the pigment components. Genetics 29(2):148–175

Landini G (2006–2020) Background illumination correction

Mendel G (1866) Versuche über Pflanzen-Hybriden. Im Verlage des Vereines. https://doi.org/10.5962/bhl.title.61004

Morgan TH (1910) Sex limited inheritance in drosophila. Science 32(812):120–122. https://doi.org/10.1126/science.32.812.120

Morgan TH (1911) The origin of five mutations in eye color in drosophila and their modes of inheritance. Science 33(849):534–537

Muller HJ (1930) Types of visible variations induced by x-rays in drosophila. J Genet 22(3):299–334. https://doi.org/10.1007/BF02984195

Sass GL, Henikoff S (1998) Comparative analysis of position-effect variegation mutations in drosophila melanogaster delineates the targets of modifiers. Genetics 148(2):733–741. https://www.ncbi.nlm.nih.gov/pmc/articles/PMC1459838/

Schindelin J, Arganda-Carreras I, Frise E, Kaynig V, Longair M, Pietzsch T, Preibisch S, Rueden C, Saalfeld S, Schmid B, et al (2012) Fiji: an open-source platform for biological-image analysis. Nature methods 9(7):676–82. https://doi.org/10.1038/nmeth.2019, citation Key: Schindelin2012

Schlaffke L, Golisch A, Haag LM, Lenz M, Heba S, Lissek S, Schmidt-Wilcke T, Eysel UT, Tegenthoff M (2015) The brain's dress code: how the dress allows to decode the neuronal pathway of an optical illusion. Cortex 73:271–275. https://doi.org/10.1016/j.cortex.2015.08.017

7

A MATLAB Pipeline for Spatiotemporal Quantification of Monolayer Cell Migration

Yishaia Zabary and Assaf Zaritsky

Contents

This Chapter has been reviewed by Simon F. Nørrelykke, ETH Zurich, Switzerland.

© The Author(s) 2022
K. Miura, N. Sladoje (eds.), *Bioimage Data Analysis Workflows–Advanced Components and Methods*, Learning Materials in Biosciences, https://doi.org/10.1007/978-3-030-76394-7_8

What You Will Learn in This Chapter

In this chapter we present a MATLAB-based computational pipeline for the quantification of monolayer migration assays. Wound healing assay (or scratch assay) is a commonly used *in vitro* assay to assess collective cell migration. Our pipeline outputs traditional and spatiotemporal readouts that quantify the group migration properties and was previously used for a screen that included thousands of time-lapse sequences. You will learn how to execute the pipeline, the principles behind the design and implementation choices we made, pitfalls, tips, and tricks in using it.

8.1 Introduction

In vitro monolayer migration assays are a simple model for studying collective cell migration, a fundamental cellular function with vast implications in health and disease. Quantification of monolayer migration is required for the investigation of the molecular and cellular mechanisms that govern collective cell migration, and is the bottleneck in many projects. Data analysis and automated quantification become absolutely essential especially due to recent advances in automated imaging-based data acquisition through high content imaging platforms, making manual annotation impractical. Wound healing (or "scratch") assay is the most common assay to quantify collective cell migration *in vitro* (Liang et al., 2007), and is performed by monitoring the "healing" of a scratch in a growing confluent monolayer of cells by still or time-lapse microscopy (Jonkman et al., 2014). The basic initial step of almost all monolayer migration analyses pipelines is the segmentation of each image to cellular and non-cellular regions. This segmentation can be then used to quantify the area covered by the monolayer in two snapshots (endpoint readout) or to calculate the rate of healing through time using live imaging (temporal readout). This type of analysis does not require fluorescent labeling and is usually performed using label-free imaging modalities such as phase contrast or differential interference contrast (DIC) microscopy. Accordingly, several open computational tools were designed to segment cellular and non-cellular image regions in label-free images with the purpose to quantify the wound healing progression (Deforet et al., 2012; Geback et al., 2009; Masuzzo et al., 2016; Milde et al., 2012; Zaritsky et al., 2017b), including several FIJI plugins (Caldas et al., 2015; Suarez-Arnedo et al., 2020) and a CellProfiler pipeline (Carpenter et al., 2006). Tracking the overall growth of confluent cell monolayers is not always sufficient to discriminate different modes of monolayer migration and to fully understand the collective dynamics. Indeed, live imaging can provide important and useful information beyond the healing rate: persistent migration (Ng et al., 2012), orientation (Milde et al., 2012; Ng et al., 2012), directional migration (Deforet et al., 2012; Milde et al., 2012; Ng et al., 2012), strain rate (Lee et al., 2013), monolayer front dynamics (Zaritsky et al., 2015b) and other measures for local or global coordinated migration (Deforet et al., 2012; Milde et al., 2012; Ng et al., 2012; Slater et al., 2013; Zaritsky et al., 2014; Zhou et al., 2019). These measures can be used to characterize and discriminate between effects of different treatments or experimental conditions (e.g. Simpson et al., 2008; Vitorino and Meyer, 2008). Importantly, valuable discriminative information can be extracted from cells within the bulk during monolayer migration (Deforet et al., 2012; Zaritsky et al., 2017b; Zhou et al., 2019), but these spatiotemporal measures are inherently less intuitive and are harder to process, analyze and interpret. In wound healing experiments with simple geometrical patterns of the monolayer front, spatiotemporal averaging of many cells based on their location relative to the monolayer's front can generate qualitative visualization and quantitative measures that can help in interpreting observed phenotypes (Zaritsky et al., 2012). In this chapter we present a computational pipeline for automated visualization and quantification that was previously robustly applied to thousands of monolayer migration experiments (Zaritsky et al., 2017b). The pipeline was implemented in MATLAB and provides the traditional "wound healing" measures as well as more advanced spatiotemporal representations that can be used for visualization as well as for quantitative analysis. The chapter contains detailed information including practical usage instructions, parameter tuning, algorithms, troubleshooting and output interpretation.

8.2 Dataset

Accessing sample data: sample data from Zaritsky et al. (2013) is available in a Zenodo dataset (Zabary and Zaritsky, 2020). The dataset is in a compressed (ZIP) file within the following folders:

- *TimeLapseSamples*—six .tif image stacks of representative time-lapse experiments for single expanding monolayers and two monolayers expanding toward each other in different cell systems and experimental conditions (■ Fig. 8.1). The specific cell system and imaging parameters are as follows:
  - The *SingleExpandingMonolayer* folder:
    - *EXP_16HBE14o_1E_SAMPLE.tif*—16HBE14o cells, imaged with pixel size of 1.267 μm and time resolution of 5 min per frame.
    - *EXP_DA3_PHA_1E_SAMPLE.tif*—DA3 cells treated with PHA, imaged with pixel size of 1.24 μm and time resolution of 14.5 min per frame.
    - *EXP_MDCK_HGFSF_1E_SAMPLE.tif*—MDCK cells treated with HGF/SF, imaged with pixel size of 0.879 μm and time resolution of 15.7 min per frame.
  - The *TwoExpandingFronts* folder:
    - *DA3_PHA_2E_SAMPLE.tif*—DA3 cells, imaged with pixel size of 1.24 μm and time resolution of 14.5 min per frame.
    - *MDCK_HGFSF_2E_SAMPLE.tif*—MDCK cells treated with HGF/SF, imaged with pixel size of 0.879 μm and time resolution of 15.7 min per frame.
    - *MDCK_ctrl_2E_SAMPLE.tif*—MDCK cells, imaged with pixel size of 0.879 μm and time resolution of 15.7 min per frame.
- *MultipleExperimentKymographs*—kymographs from multiple time-lapse experiments for post-processing analysis.

8.2.1 Experimental Considerations

The pipeline supports two experimental settings: (1) A single expanding monolayer and (2) the standard "wound healing" or "scratch" assay that includes two monolayers

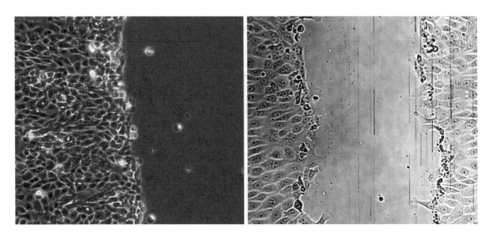

■ **Fig. 8.1** **Experimental setup.** Our pipeline supports experiments of single expanding monolayers (left) or two monolayers expanding toward each other (right). Left: a single expanding monolayer of Human bronchial epithelial cell monolayer (16HBE14o line) live imaged with phase contrast micrsocopy. Right: the traditional wound healing (or scratch) assay, two monolayers of Madin-Darby Canine Kidney (MDCK) cells live imaged with differential interference contrast (DIC) microscopy. For spatiotemrpoal quantification we recommend the single expanding monolayer setting (see text). Images from Zaritsky et al. (2015a)

advancing toward one another. Examples are shown in ■ Fig. 8.1. We recommend performing experiments with a single expanding monolayer for two reasons. First, the segmentation algorithm implemented in this pipeline performs less accurately once the two advancing monolayers are very close. Second, our spatiotemporal analysis is based on having sufficient bulk of cells behind the monolayer advancing front (for more spatial information) together with sufficient free space for the monolayer to expand (for more temporal information), both are easier to meet with single expanding monolayer experiments. This is especially true given that most studies do not use the information following the monolayers collision. For simplicity, we will use the term "wound healing" for both settings. Note that the pipeline supports monolayers expanding in the horizontal axis (x-axis). Thus, a 90° rotation preprocessing step (e.g. with FIJI) is required before analyzing vertically expanding monolayers.

8.3 Tools

The custom analysis pipeline was implemented in MATLAB. You must have a MAT-LAB license in order to use the pipeline. Note that many academic institutions have campus-wide MATLAB licenses, contact the IT in your institution for details. The pipeline was tested with MATLAB version 2019b on Mac/Windows/Linux operating systems, it will not function properly on earlier MATLAB versions. The pipeline produces multiple types of outputs, ranging from the standard "healing rate" to advanced spatiotemporal visualizations and quantifications (see ■ Fig 8.2). Note, that the raw experimental data consists of live label-free imaging, and thus the pipeline is based on Particle Image Velocimetry (PIV) (Santiago et al., 1998), rather than single cell

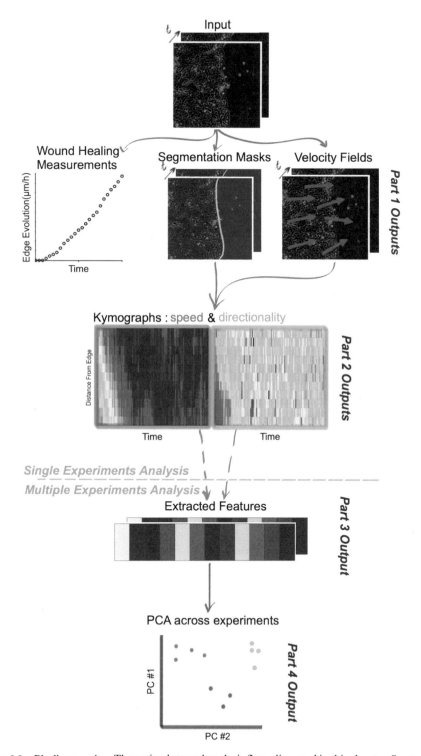

◨ **Fig. 8.2** **Pipeline overview:** The entire data and analysis flow, discussed in this chapter. See text for full details

tracking. In the coming sections of this chapter we made efforts to provide detailed explanations for users inexperienced in MATLAB programming, as well as a modular and documented implementation to enable flexibility and customization for experienced users.

8.3.1 Setting up the MATLAB Environment and Executing the Analysis Pipeline

All the analysis pipeline code will be placed in the *working directory*.
All the raw data and the outputs will be placed in the *data directory*.

1. Accessing the code: the complete code is available in a GitHub repository https://github.com/assafZaritskyLab/SpatiotemproalQuantificationMonolayerCell MigrationPipeline.git The repository includes MATLAB source code and a sample dataset.
2. To analyse a single time-lapse, use the script *quantifyMonolayerMigrationMain.m*; for convenience we shall use the abbreviation *main.m*. The script requests as input a label-free image stack (the supported formats are .tiff/.zvi/.lsm), and a minimal set of parameters (the physical pixel size in μm, the time resolution in minutes, one/two expanding monolayers, the cell line's approximate maximal speed and two more parameters for the time interval of the analysis). The script executes the analysis with the default parameters, and takes as default the 1st channel of a multi-channel data.
3. To analyse a set of experiments use the script *quantifyMonolayerMigrationBulk-Main.m*. For convenience we will use the abbreviation *mainBulk.m*. The script is a batch-processing version of *main.m* for multiple image stacks. It requests as input a path to a directory containing multiple label-free image stacks (supporting the same formats) and the same set of mandatory parameters. The script executes the analysis for each label-free image stack using the default parameters, followed by "meta analysis", extracting information from the complete dataset (for advanced users, detailed information can be found later).

8.4 Workflow

In the following sub-sections we will provide a detailed description of the analysis pipeline. After reading it you will be able to understand the input/output of each step, tune parameters and troubleshoot the execution on your data.

8.4.1 Pipeline Overview

The pipeline receives as input the raw image data in one of the following formats: *tiff* stack, *zvi* (Zeiss Vision Image) or *lsm* (Zeiss tiff based proprietary format). Each data file is a single time-lapse experiment. The overview of the pipeline is presented in ◻ Fig. 8.2.

The pipeline includes four conceptual steps, which are reflected in four central MATLAB functions, each depending on the previous one and thus must be executed sequentially (see lines 88–104 in *mainBulk.m* script). The first two steps are performed at the single time-lapse level (see lines 74–79 *main.m* script). The rest of the pipeline is for the analysis of multiple experiments, enabling the comparison between different experiments and conditions, and might be challenging for novice users (see *mainBulk.m* script).

Part 1: Segmenting each image to cellular (foreground) and background regions, and calculating the velocity fields. The output of this stage includes quantification of the wound healing over time, visualizations of the foreground/background segmentation, visualization of the velocity fields, and more detailed visualization of outputs for advanced debugging purposes (◘ Fig. 8.2, 2nd row). The functions that produce each of these outputs are invoked by the function *StepsScripts/calcSpatiotemporalRaw.m* lines 25–30.

Part 2: Calculating kymographs that capture the experiment's spatiotemporal dynamics. The output of this stage includes visualization of the kymographs (◘ Fig. 8.2, 3rd row) and is generated by the function *StepsScripts/KymographsByMeasure.m*.

Part 3: Extracting spatiotemporal feature vectors from each kymograph (◘ Fig. 8.2, 4th row) using the *StepsScripts/kymographToFeaturesVec.m* function on each single experiment.

Part 4: Calculating the principal components of these features across experiments (◘ Fig. 8.2, 5th row). This is performed with the *StepsScripts/PCAOnAllExperimentsMeasurements.m* function, which uses MATLAB's built-in pca (2020) function.

Parameter Initialization

There are seven parameters that must be explicitly specified by the user: (1) the physical pixel size (in μm), (2) the time resolution (in minutes), (3) the advancing monolayer estimated maximal speed (in $\mu m * h^{-1}$), (4) the number of expanding monolayers (one or two), (5–6) the time interval: initial and final frame for the analysis, (7) the patch size (in μm, explained later). Other parameters can be set manually withing the code, for example, *reuse* enables or disables the use of past calculations (see line 3 in *main.m*). Default parameters are set in *utils/initParamsDirs.m* (invoked in *main.m*, line 73).

◘ Table 8.1 summarizes the parameters that can be adjusted by the user via the GUI. The exact purpose of each parameter and the effect of altering parameters will be discussed later.

Output Directory Structure

The pipeline outputs will be automatically generated and placed in the data directory.

1. The outputs of each time lapse sequence will be located in a directory named according to the raw data file name that contains the following sub-directories: (1) '*images/*' a separate raw image file for each frame in the time lapse sequence, (2) "*VF/*" debug outputs and results for the velocity fields analysis, and (3) "*ROI/*" debug outputs and segmentation results.

◻ Table 8.1 Main parameters used in the pipeline

Parameter name	Unit of measure	Required as user input	Default value	Description
pixelSize	μm	Yes	1.267428	Physical pixel size
timePer-Frame	minutes	Yes	5	Time between consecutive frames (temporal resolution).
nRois	–	Yes	1	The number of expanding monolayers (1 or 2). The parameter nRois stands for the number of regions of interest (ROI)
maxSpeed	$\mu m/h$	Yes	90	The estimated maximal speed of the inspected phenotype.
minN-Frames	–	Yes	1	The index of the first frame to include in the analysis.
maxN-Frames	–	Yes	Total Number of frames	The frame number to end the analysis on.
patchSize	μm	Yes	15	The patch size for PIV and kymograph analyses

2. The outputs that relate to a full time lapse experiment are located in designated directories that are generated in the data directory. For example, plots of the wound healing rate over time for all experiments will be located in a designated directory to allow straightforward comparison between different experiments. These directories are: (1) *"segmentation/"* videos with visualization of the segmentation results, (2) *"monolayerMigrationMeasures/"* plots and data relating to the wound healing readouts, (3) *"kymographs/"*, spatiotemporal quantification and visualization of the experiment, (4) *"kymographFeatures/"*, quantitative features extracted from the kymographs, and (5) *"PCA_Results/"*, dimensionality reduction results.

From here, each of the four steps of the analysis are explained in detail.

8.4.2 Part 1: Estimation of Velocity Fields, Semantic Segmentation, and Calculation of Wound Healing Measurements

This part starts with the raw image data, calculating the velocity fields followed by segmentation of the foreground cellular regions in each image and includes a correction for microscope re-positioning error. This step is implemented by the function *StepsS-*

cripts/calcSpatiotemporalRaw.m. The output of this stage includes quantification of the wound healing rate over time, and visualizations of the foreground/background segmentation and velocity fields. This part also provides detailed visualization of the output in every frame for troubleshooting and debugging.

Estimating Velocity Fields

We start by estimating the velocity fields for each frame in the time-lapse sequence. Velocity fields were computed using custom cross-correlation-based particle image velocimetry (PIV), utilizing non-overlapping image patches. This is illustrated in ◻ Fig. 8.3a, and implemented in *utils/whscripts/whLocalMotionEstimation.m* function (below).

The frame-to-frame displacement of each patch was defined based on the maximal cross-correlation of a given patch with the subsequent image in the time-lapse image sequence.

```
41   [dydx, dys, dxs, scores] = blockMatching(I0, I1,
     ↪    params.patchSize,params.searchRadiusInPixels,true(size(I0)));
     ↪    % block width, search radius,
```

utils/whScripts/whLocalMotionEstimation.m:41

The search radius was constrained by the `searchRadiusInPixels` parameter that was set based on the estimated maximal cell speed and the temporal resolution:

```
85   params.searchRadiusInPixels = ceil((params.maxSpeed/params.pixelSize)
     ↪    * (params.timePerFrame*params.frameJump/60));
```

utils/initParamsDirs.m:85-87

Processing time is dependent on the experiment data (size in pixels of each frame) and parameters (such as cell maximal speed and patch size). Processing of a single frame in the sample data with the experiment-specific default parameters may take up to 4 s on a standard laptop.

Segmenting the Cellular Foreground

For each frame in the time-lapse sequence, each patch is assigned as foreground or background, and this binary classification (segmentation) is used to calculate the contour of the migrating monolayer. The segmentation algorithm relies on two priors: (1) each frame contains one/two continuous "cellular foreground" segment/s and one continuous "background" segment, and (2) the contour advances monotonically over time toward the empty space. These assumptions allow us to compute the initial contour at time 0, as implemented in the custom function which takes as input the image I, the `patchSize`, and `lbpMapping` (an internal structure calculated beforehand and required for the segmentation), and outputs the segmentation mask `roiTexture`.

Then, we use the segmentation at time t as a seed to expand the "cellular foreground" to time $t + 1$. The only patches to be resolved at time $t + 1$ are those labeled

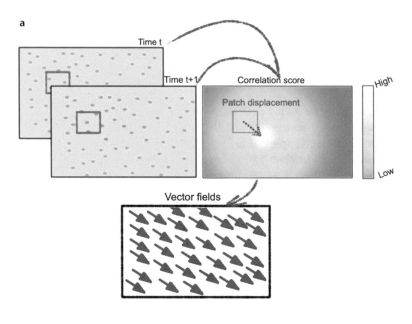

8

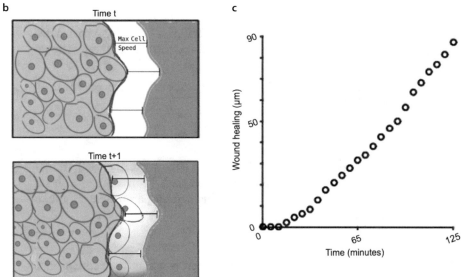

□ **Fig. 8.3   Part 1: velocity fields, segmentation and wound healing measurements.** (a) Particle Image Velocimetry (PIV). Depiction of velocity estimation for a patch (blue square). The maximal correlation is calculated between the patch at time t and all potential translations in frame $t + 1$, and the corresponding velocity vector is recorded. (b) The segmentation problem is reduced to a narrow band that is defined based on the current contour (blue) and the maximal cell speed (green). (c) Wound healing plot, calculated using the segmentation masks

```
85    roiTexture =
      ↪    segmentPhaseContrastLBPKmeans(I,params.patchSize,lbpMapping);
```

segmentation/temporalBasedSegmentation.m:85.

as "background" at time t and within a cell motion reach in respect to the monolayer contour (based on maxSpeed); this is illustrated in ◼ Fig. 8.3b. The function

```
105   ROI = (curRoi & dilate(prevRoi,changeRadius));
```

segmentation/temporalBasedSegmentation.m:105.

computes ROI, a binary mask of the estimated cellular foreground at time $t + 1$. curRoi is calculated from thresholding the PIV cross-correlation scores followed by morphological operators.

Calculating the Wound Healing Over Time

The wound healing can be calculated as the expansion (in μm) of the monolayer front over time:

```
52    nDiffPixels = sum(ROI1(:)) - sum(ROI0(:)); % diff from previous frame
53    if t == 1
54        healingUm(t) = params.pixelSize * nDiffPixels /
          ↪    monolayerAbsWidth;
55    else
56        healingUm(t) = healingUm(t - 1) + params.pixelSize * nDiffPixels
      ↪    / monolayerAbsWidth;
57    end
```

StepsScripts/CalcMonolayerMigrationMeasures.m:52-53.

ROI1 is the segmentation mask at the given time point, ROI0 is the segmentation mask at the previous time point. HealingUm(t) is the accumulated edge expansion at time t. This is illustrated in ◼ Fig. 8.3c.

The wound healing rate is calculated as the instantaneous or the average change in the wound healing over time (temporal derivative) in $\mu m * hour^{-1}$:

```
56    healingRate(t) = params.toMuPerHour * nDiffPixels / size(ROI0,1);
57    averageHealingRate(t) = params.toMuPerHour * nDiffPixelsMeta /
      ↪    (size(ROI0,1) * t);
```

StepsScripts/CalcMonolayerMigrationMeasures.m:56-57

Part 1: Outputs

The execution of each part in the analysis pipeline is dependent on the successful execution of the previous steps. Thus, each part generates outputs (.mat format) to be used as input for the following step/s, as well as outputs for quantification, visualization and debugging (in multiple formats). ◨ Table 8.2 contains full description of the outputs of Part 1 of the pipeline.

◨ **Table 8.2 Outputs of Part 1 of the pipeline.** 'MATLAB function'—the function that generates the output and is available in the source code for independent use. Ôutput directory'—path to the outputs, the root directory is the working directory (see ▶ Sect. 8.4.1 for the output directory structure). 'File format'—the output filetypes: *.mat* data files, *.eps*, *.fig* are vector graphics and *.jpg* are image files, *.avi* files are video files

Output	MATLAB function	Output directory	File format	Brief description
Velocity fields	*Estimate VelocityFields*	expPrefix/VF/vf expPrefix/VF/vfOrig/	.mat	The directory ' VF' contains the velocity fields of each frame In 'vfOrig/' before correction for microscopy repositioning errors, and in 'vf/' after the the corrections (used for further processing)
Segmentation mask	*temporalBased Segmentation*	expPrefix/ROI/roi/	.mat	The segmentation mask for each frame
Segmentation movie	*SegmentationMovie*	segmentation/	.avi	A video that visualizes the image sequence with the segmented contour overlaid
Wound healing readouts	*CalcMonolayer MigrationMeasures*	monolayerMigration Measures/	.mat .eps .jpg .fig .csv	Quantification and visualization of the wound healing and its rate
Velocity fields visualization	*renderVelocity FieldVideo*	expPrefix/VF/vfVis/	.eps .jpg .fig .avi	Velocity fields visualization, raw images with the velocity fields overlaid in separate files or as a video

Part 1: Parameter Sensitivity and Trade-Offs

The two parameters that have the most influence on the velocity fields calculation are the patch size and the cells' maximal speed, as illustrated in ■ Figs. 8.4 and ■ 8.5. Patches that are too small do not contain sufficient image texture to statistically establish the optimal translation to the next time-frame, leading to spatial inconsistencies in the velocity fields. On the other hand, patches that are too large may include texture from multiple entities that move in different directions, leading to conflicting local motion patterns within the patch and impairing coherent motion estimation. An example is shown in ■ Fig. 8.4b, with patch size equal to 30 μm. Other considerations include patch-size dependent (quadratic) velocity fields calculation time as patches decrease, and reduction in the resulting spatial resolution in a (quadratic) patch-size dependent manner. These inherent trade-offs are optimized by selecting a patch size smaller than the size of an average cell, and visually validating the coherency and resolution of the velocity fields outputs. A second validation that relies on the resulting kymographs will be discussed in the description of Part 2 of the pipeline. From our experience, a patch size of 15 μm performs well for several cell lines and microscopy objectives.

The parameter related to the cell's maximal speed is used to determine the search radius for the PIV calculations, where larger maximal speed requires a larger search radius. The actual value for this parameter is dependent on cell type and exact experimental setting, and should be determined based on the data. A too small search radius will lead to underestimation of the velocity fields magnitude (illustrated in ■ Fig. 8.4b, for max speed equal to 10 $\mu m * hr^{-1}$). A search radius beyond the one defined based on the true maximal cell speed will lead to more errors in the velocity field estimation, due to the quadratic increase in the number of possible translations. For example, detection of high motility patches in the background is presented in ■ Fig. 8.4b, where the background speed for 200 $\mu m * hr^{-1}$ and for 90 $\mu m * hr^{-1}$ are to be compared, with patch size of 15 μm. In addition, the execution time is quadratic in the search radius size. Note that the pipeline is not very sensitive to the value of this parameter. Our recommendation is to assign this parameter using prior knowledge regarding the cell system and the experiment and validate visually using the resulting velocity fields (and/or kymographs, as we will discuss in Part 2).

The segmentation algorithm was optimized for robust high-content automated analysis. To achieve robust segmentation we relied on the assumption that the image includes one/two continuous monotonically expanding monolayers, thus dramatically reducing the number of patches to be resolved in each iteration (as shown in ■ Fig. 8.3b, green). Active contours and graph cut algorithms in general produced accurate segmentation masks that were not limited by the patch size resolution (Zaritsky et al., 2011), however these approaches also led to large errors in segmentation of some image frames. We have observed that small segmentation inaccuracies have a minor effect on the wound healing measurements, as well as on the kymographs (Part 2), whereas large segmentation errors, even if only in one frame in a time lapse, may cause major artifacts in the calculated readotus (especially in the kymographs). Thus, we compromise for slightly reduced (overall) segmentation accuracy, to achieve enhanced robustness. This approach was found to be very effective in a large screen, allowing us to exclude less than 1% of the experiments due to major segmentation errors (Zaritsky et al., 2017a). Our recommendation is to perform a visual validation of the segmentation outcome. In case of identifying an erroneous frame, its corre-

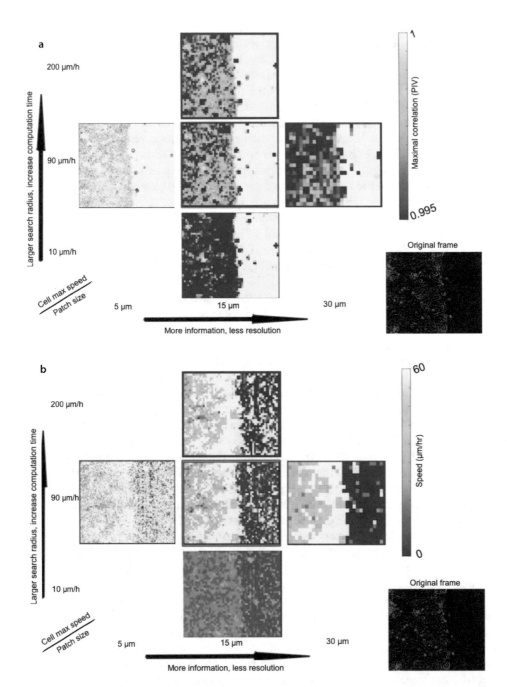

■ **Fig. 8.4 Parameters trade-offs in PIV calculation.** The maximal correlation (**a**) and the estimated speed (**b**), as a function of the parameters for patch size (x-axis) and maximal cell speed (y-axis). The middle frame (red boundaries) corresponds to the pipeline's default parameters. The image frame used in this figure is shown at the lower-right corner

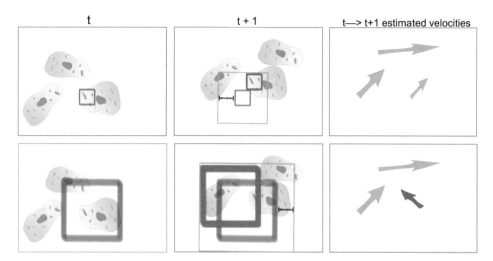

◻ Fig. 8.5 Patch size sensitivity and trade-offs. A cartoon illustrating how PIV is affected by the patch size. Subcellular (top row) versus supercellular (bottom row) patch size. The patch is marked in dark blue, and the search radius, which is calculated from the cell's maximal speed and time resolution (black delimiter), is marked in orange. Columns represent (left-to-right) frame t, frame $t + 1$ and estimated velocities. The estimated displacement of the patch is marked in pink (second column). (Impaired displacement calculation for supercellular size patches containing texture from multiple cells is shown in the bottom-right, as red arrow)

sponding segmentation mask (ROI) can be replaced with a previous or following frame without major effects (◻ Table 8.3).

Part 1: Practical Usage of the Outputs

The outputs of Part 1 include the traditional wound healing readouts of the wound healing over time and the wound healing rate. These measurements can be compared across experiments and treatments. The visualizations can be used for setting and validating parameter values (as discussed above, in the the previous subsection). The visualization of the segmentation masks is important to verify that the segmentation follows the evolving contour, key for proper quantification of the wound healing readouts and spatiotemporal quantification (which we will discuss later in this Chapter)

? Exercise 1

Several algorithms were proposed for monolayer segmentation. For example, Geback et al. (2009) used discrete curvelet transform, wheras Candès et al. (2006) and Zaritsky et al. (2011) used Support Vector Machine and Graph Cuts. These algorithms usually produce segmentation masks with higher accuracy, in comparison to the algorithm used here. Explain what was the reason for the algorithmic design choice in this pipeline.

◻ **Table 8.3** Parameters used in Part 1 of the pipeline

Parameter name	Unit of measure	Required as user input	Default value	Description
maxSpeed	μm/h	No	90 μm/h	The maximal expected cell speed. Used to define the search radius for the Particle Image Velocimetry.
patch-SizeUm	μm	No	15 μm	Patch size for the Particle Image Velocimetry We recommend setting this value to be smaller than the cell diameter.
searchRadiusInPixels	–	No	$\frac{\frac{maxSpeed}{pixelSize} * timePerFrame*frameJump}{60}$	Search radius for the Particle image velocimetry Default value calculated based on the previous parameters.
maxNFrames	–	No	Number of frames in the time-lapse experiment	The number of frames to be analyzed.

8.4.3 Part 2: Kymographs

The spatiotemporal dynamics of a full time-lapse experiment can be quantified and visualized in kymographs. At each time point, the distance from the monolayer front was calculated using the segmentation masks, and the velocity fields were used to measure the cells migration properties at a given time and location with respect to the front. More specifically, in each frame, the cellular foreground is divided into bands of constant distances from the monolayer front, termed *strips*. Each bin in the kymographs records the cells' mean speed or directionality in a specific strip at a particular time point. This is illustrated in ◻ Fig. 8.6. Speed is calculated as the magnitude of the corresponding velocity vector, while directionality is the absolute

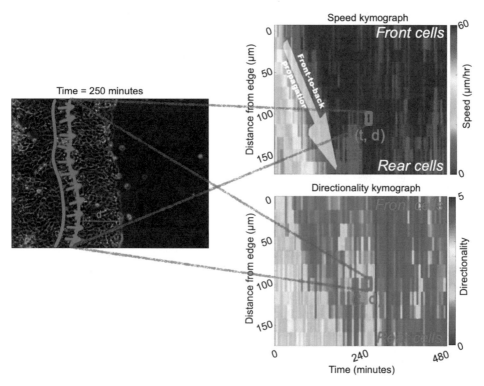

◘ Fig. 8.6 Kymograph construction. Speed (top right) and directionality (bottom right) kymographs provide a compact representation of the complete time-lapse sequence. Each bin (t,d) holds the average speed and directionality (accordingly) of all patches at time t and distance d from the monolayer front

ratio between the mean velocity component perpendicular to the monolayer front and the velocity component parallel to the monolayer front:

```
inDist = (DIST > (params.strips(d)-params.kymoResolution.stripSize))
↪    & (DIST < params.strips(d)) & ~isnan(speed);
speedInStrip = speed(inDist);
```

Calculation of a single bin in a speed kymograph, utils/whScripts/whKymograph.m:73-74.

`DIST` is an image where each pixel encodes its Euclidean distance from the monolayer front. `inDist` is a binary mask of all pixels within a specific strip in index d.

Part 2: Outputs

The outputs of Part 2 of the pipleine are speed and directionality kymographs for visualization and further analysis; they will be passed on to Part 3. ◘ Table 8.4 contains their detailed description.

◨ **Table 8.4**   Outputs of Part 2 of the pipeline

Output	MATLAB function	Output directory	File format	Brief description
Speed kymo-graph	*whKymographs*	kymographs/	.mat .fig .jpg .eps	Quantification and visualization of the speed kymograph.
Directionality kymograph	*whKymographs*	kymographs/	.mat .fig .jpg .eps	Quantification and visualization of the directionality kymograph.

Part 2: Parameter Sensitivity and Trade-Offs

As the kymographs are calculated from the results of the previous analysis steps, potential errors in calculations in Part 1 will lead to inaccuracies in the kymographs. For example, if the search radius is set to a value smaller than the actual cell speed, the resulting vector field magnitudes and the kymograph values will be lower than the actual velocities, as illustrated in ◨ Fig. 8.7 (left). When the search radius is much higher than the maximal cell speed, the potential matching translations for each patch grow quadratically, leading to over-estimation of the velocities. ◨ Fig. 8.7 (right), illustrates this situation. Faulty segmentation can also lead to erroneous kymographs by altering the bands in relation to the monolayer front, as shown in ◨ Fig. 8.8.

Four parameters control the spatial and temporal ranges for which the kymograph is calculated for; this is illustrated in ◨ Fig. 8.9. The parameters include `kymoMinDistMu` and `kymoMaxDistMu` that define the spatial region in relation to the monolayer front, and `kymoMinTimeFrameNum` and `kymoMaxTimeFrameNum` that define the temporal range for kymograph calculation (◨ Fig. 8.9). These parameters are used to calculate the internal parameter *strips*, an array of masks for each *strip* in the cellular foreground, allowing for fast retrieval of all velocity fields within each *strip*:

```
params.strips =  ceil(params.kymoMinDistMu/params.pixelSize) :
  ↪   params.kymoResolution.stripSize : params.kymoResolution.max;
```

utils/initParamsDirs.m:138-139

The purpose of setting these parameters is to enable focusing the spatiotemporal visualization and quantification to specific regions of interest in space and time. For example, if the research question relates exclusively to cells deep within the bulk then the spatial parameters can be set such that the range `kymoMinDistMu` to `kymoMaxDistMu` captures these cells of interest (◨ Table 8.5).

Part 2: Practical Usage of the Outputs

Kymographs can serve for visualization (e.g., Gan et al., 2016) and/or quantification, enabling comparison of the effect of different experimental conditions on spatiotem-

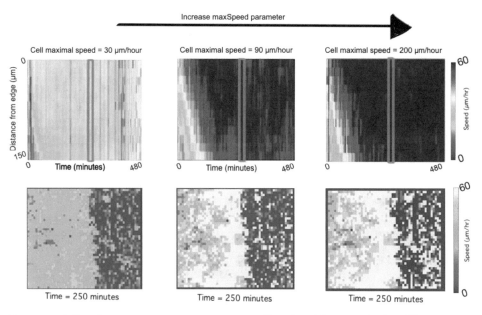

Fig. 8.7 Effect of the search radius on the kymographs. Top: Speed kymographs for different search radius values. The search radius is determined by the cell's maximal speed parameter. Underestimation (left) and (minor) overestimation (right) of cell speed due to low and high maximal cell speed values, correspondingly. Bottom: Speed snapshots (at time = 250 min), corresponding to the magenta vertical band in the kymograph directly above

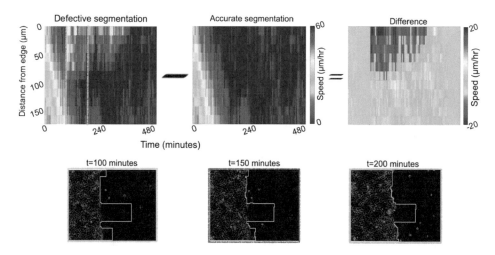

Fig. 8.8 Effect of faulty segmentation on the kymograph. The top-left panel shows the speed kymographs altered by the defective segmentation. The colored (cyan, blue, green) kymographs's columns represent the corresponding corrupted segmentation of individual frames (bottom). Importantly, the segmentation algorithm considers these faulty frames as the segmentation which includes a temporal continuity assumption (see Part 1). The top panel visualizes the deviation of the kymograph caused by the corrupted segmentations. The kymograph on the right is the subtraction of the kymograph (middle) from the defected kymograph (left)

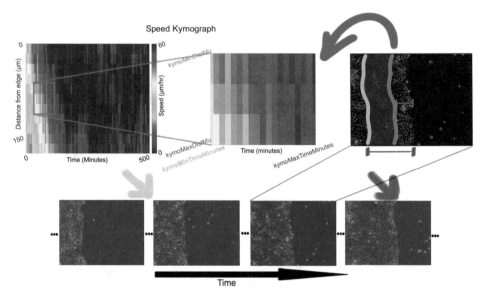

◘ Fig. 8.9   Controlling the kymograph's spatial and temporal range. Top left: A speed kymograph (using the pipeline's default parameters, see ◘ Table 8.5). Top middle: The kymograph calculated with reduced temporal and spatial ranges (kymoMinDistMu = 60 μm, kymoMaxDistMu = 105 μm, kymoMinTimeMinutes = 40 min, kymoMinTimeMinutes = 130 min). Bottom: Snapshots from the time-lapse sequence, arrows point to the kymoMinTimeMinutes (cyan) and to the kymoMaxTimeMinutes (red). Top right: The spatial region defined by the kymoMinDistMu (orange) and the kymoMaxDistMu (purple) parameters

◘ Table 8.5   Parameters used in Part 2 of the pipeline

Parameter name	Unit of measure	Required as user input	Default value	Description
kymo-MinDistMu	μm	No	patchSize	Minimal distance from the monolayer front used for kymograph construction (◘ Fig. 8.9)
kymo-MaxDistMu	μm	No	180	Maximal distance from the monolayer front used for kymograph construction (◘ Fig. 8.9)
kymoM-inTimeMinutes	minutes	No	0	Minimal time used for kymograph construction (◘ Fig. 8.9)
kymoMax-TimeMinutes	minutes	No	0	Maximal time used for kymograph construction (◘ Fig. 8.9)

poral monolayer migration dynamics. For example, in Zaritsky et al. (2017b), we used the kymographs to visualize an overall motility impairment following inhibition of some proteins, and to reveal a rapid front-to-back motility synchronization, as a response to inhibition of a specific pathway, that could not be discovered without systematic spatiotemporal visualization and quantification. This is illustrated in ◘ Fig. 8.10. The kymographs can also be used as indication for a successful or defective analysis. For example, if most values in the directionality kymograph are close to 0, that might indicate that the monolayer advances vertically and should be rotated to advance horizontally (more examples follow in the section on parameters sensitivity and trade-offs). In Parts 3 and 4 we use the kymographs as high-dimensional quantitative readouts.

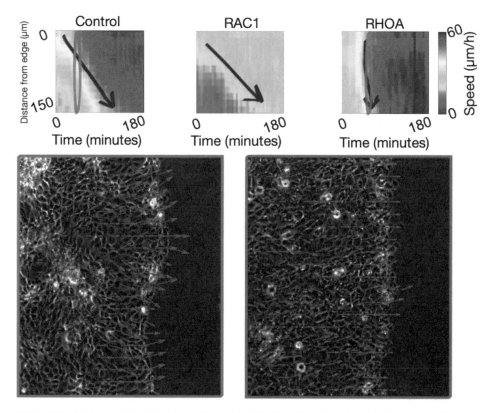

◘ **Fig. 8.10 Kymograph visualization enables new insight.** Top: Speed kymographs for a control experiment (left), RAC1 inhibited cells (middle) and for RHOA inhibited cells (right). RAC1 inhibition leads to reduced motility in space and time. The steeper arrow indicates a faster front-to-back motility propagation for the RHOA depleted cells. Bottom: A snapshot of the control (left) and RHOA inhibited experiment (right) (time = 60 min, border color code matches the corresponding kymograph's column) with overlaid velocity fields. The velocities of the RHOA treated cells at the front and the back of the monolayer are more synchronized than the control cells. Adapted from Zaritsky et al. (2017b)

? Exercise 2

The velocity fields estimation is sensitive to the values assigned to the parameters `patchSizeUm` and `maxSpeed`. In this and the following exercises you will explore how alteration in these parameters affects the resulting kymographs.

Download the image sequences *EXP_16HBE14o_1E_SAMPLE.tif* and *EXP_MDCK_HGFSF_1E_SAMPLE.tif* (see ▶ Sect. 8.2). Calculate and visually compare their speed and directionality kymographs under the following parameter configurations:

1. `patchSizeUm` = 15 μm, `maxSpeed` = 90 $\mu m * h^{-1}$, being the default values of the parameters;
2. `patchSizeUm` = 5 μm, `maxSpeed` = 30 $\mu m * h^{-1}$.

Discuss the obtained results.

? Exercise 3

Visualize the kymographs computed in Exercise 2, utilizing the function `utils/plotKymograph.m`. Describe and explain the effect that the different parameters have on the resulting kymographs.

? Exercise 4

The traditional analysis of wound healing experiments includes measurement of the wound healing rate, the change in the monolayer's front evolution over time. Think of, and describe scenarios where, upon a perturbation, the wound healing rate remains unchanged, but other collective migration properties change.

? Exercise 5

Visualizing the subtraction of two kymographs can provide valuable insights regarding the corresponding differences in their spatiotemporal dynamics. Write a code snippet to compute the subtraction of the two speed kymographs that you obtained in Exercise 2, for the case when the default parameters are used. Visualize the results.

8.4.4 Part 3: Feature Extraction

This part of the analysis pipeline compresses the kymograph to a feature vector as a more compact representation of the monolayer's spatiotemporal dynamics. This is achieved by averaging the bins of a kymograph in space and time and starts by dividing the kymograph into a grid of (*timePartition* x *spatialPartition*) tiles:

```
57   nFeats = params.timePartition * params.spatialPartition;
58   iSpace = 1 : singleTileSpace : nSpace+1;
59   iTime = 1 : singleTileTime : nTime+1;
```

StepsScripts/kymographToFeaturesVec.m:57-59.

The kymograph partition is defined by the indices iSpace (spatial partition) and iTime (temporal partition).

After that, each feature is computed as the mean value of all the kymograph bins that reside in the corresponding tiles (as illustrated in ◘ Fig. 8.11):

```
66  values = kymograph(ys(y):(ys(y+1)-1),xs(x):(xs(x+1)-1));
67  values = values(~isinf(values));
68  values = values(~isnan(values));
69  features(curFeatI) = mean(values(:));
```

StepsScripts/kymographToFeaturesVec.m:66-69.

This creates a feature vector as a compressed representation of the spatiotemporal information encoded in the kymograph. In an example shown in ◘ Fig. 8.11, we use timePartition = 4 and spatialPartition = 3, inducing a 12-dimensional feature vector, where features 1–4 encode the acceleration of cells at the monolayer front, features 5–8 encode the acceleration of cells 50–100 μm behind the monolayer front, feature 1, 5, and 9 encode the spatial variations in speed at the onset of the experiment, and features 3, 7, and 11 encode the spatial variation at later times. This representation was first described in Zaritsky et al. (2012).

Part 3: Outputs (See ◘ Table 8.6)

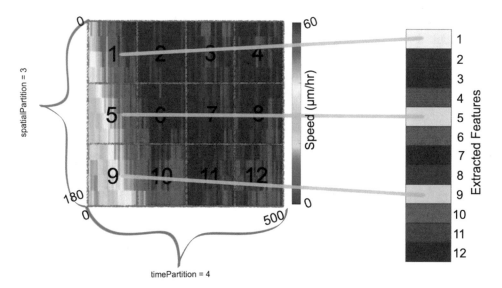

timePartition = 4

◘ **Fig. 8.11** **Reducing the representation of the spatiotemporal dynamics to a multi-dimensional feature vector.** To obtain a compact representation of a speed kymograph, we average it across space and time to *timePartition* x *spatialPartition* features. Each feature (right) encodes the average speed (color) in the corresponding kymograph's bins (left). The same process is applied to directionality kymographs

■ **Table 8.6**   Outputs of Part 3 of the pipeline

Output	MATLAB function	Output directory	File format	Brief description
Extracted features for each experiment	*kymographsTo FeaturesExtractor*	kymograph Features/	.mat	Feature vector of the corresponding time-lapse experiment

Part 3: Parameter Sensitivity and Trade-Offs

The two parameters used in Part 3 are timePartition (default = 4) and spatialPartition (default = 3), the number of temporal and spatial bins correspondingly (as illustrated in ■ Fig. 8.11). Larger values of these parameters lead to more features, smaller values to a more compressed representation. For example, for timePatition = 2, each feature will encode the temporal information for half of the experiment's duration, the same applies for the spatial component (■ Table 8.7).

Part 3: Practical Usage of the Outputs

Having a compact quantitative representation for a time-lapse experiment is key for systematic spatiotemporal statistical characterization and quantification of high-content and screening projects, where barely staring at kymographs and describing them is not sufficient (or feasible). Note that Part 3 is an intermediate step, and is usually followed by supervised (classification) or unsupervised (clustering, dimensionality reduction—see Part 4) machine learning that take high-dimensional feature vectors as input.

■ **Table 8.7**   Part 3 parameters

Parameter name	Unit of measure	Required as user input	Default value	Description
timePartition	–	No	4	Temporal partitioning used for feature extraction (■ Fig. 8.11)
spatialPartition	–	No	3	Spatial partitioning used for feature extraction (■ Fig. 8.11)

8.4.5 Part 4: Principal Component Analysis: PCA

Principal Components Analysis (PCA) is a dimensionality-reduction method transforming high-dimensional data sets to a set of individual linearly uncorrelated (orthogonal) dimensions (called Principal Components, or PCs), while preserving most of the variability in the data (Pearson, 1901). Such dimensionality reduction is performed in Part 4 of the pipeline, which receives a set of kymograph-extracted features (Part 3) from multiple experiments, normalizes the features and transforms them to a new representation of PCs, ranked by the variability that they explain:

```
94  normalized_value = (singleFeatureStruct(i) - featMean)/featStd;
95  normalizedFeatures(i) = normalized_value(1);
```

StepsScripts/PCAOnAllExperimentsMeasurements.m:94-95.

```
86  [coeff,score,latent] = pca(normalizedSingleMeasureFeatures);
```

StepsScripts/PCAOnAllExperimentsMeasurements.m:86.

Here, `normalizedSingleMeasureFeatures` is the normalized vector of either the speed or directionality kymograph features, and `pca 2020` is a MATLAB built-in function.

The obtained PCs can be used to visualize, quantify, and sometimes interpret spatiotemporal alterations between different experimental conditions (Zaritsky et al., 2012, 2017b).

Part 4: Outputs (See ◻ Table 8.8)

◻ **Table 8.8** Outputs of Part 4 of the pipeline

Output	MATLAB function	Output directory	File format	Brief description
PCA score, coefficients and latent	*PCAOnAll Experiments Measurements*	PCA_Results/	.mat	scores—the transformed PCs coefficients—the coefficients for calculating the principal components latent—PCs variances PCs are in descending order of component variance

Part 4: Practical Usage of the Outputs

Each PC can be used as a quantitative readout of the monolayer migration's spatiotemporal dynamics. By focusing on the few first PCs that capture most of the variability in the data, one can visualize, cluster and interpret to distinguish between different experimental conditions / treatments, as illustrated in ◘ Fig. 8.12. To demonstrate the processing of multiple experiments, we provide a small dataset of previously computed kymographs (in the *MultipleExperimentKymographs/* directory) and the code snippets bellow.

The function *getLabelsAndPaths* retrieves the experiments' labels and paths to speed (or directionality) kymographs:

```
measure = 'speed';
[experimentsLabels, kymographPaths] =
    getLabelsAndPaths(pathToKymographsFolder, measure);
```

Note that `getLabelsAndPaths` is a custom function that is dependent on a specific file-organization scheme. It was implemented such that each sub-directory

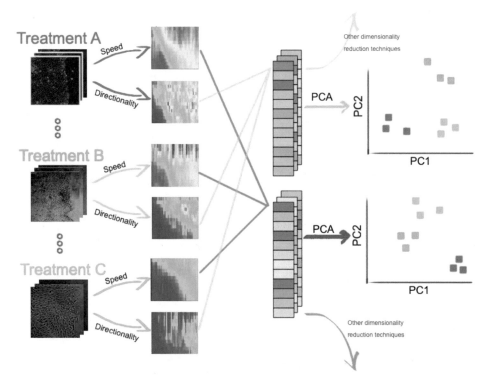

◘ **Fig. 8.12** **Depiction of the analysis of multiple experiments.** Time-lapse sequences (left column) are the raw data used to calculate the speed and directionality kymographs (second column from the left), which are then used to calculate the monolayer's spatiotemporal dynamics features vectors (third column), that are further compressed with PCA for visualization or quantification (rightmost column)

within `pathToKymographFolder` contains a collection of kymograph of either the speed or directionality measurements, named according to the experiment's name (see ▶ Sect. 8.4.1). For a different directory arrangement or labeling scheme, the `getLabelsAndPaths` function should be re-implemented accordingly.

Next, we set the necessary parameters (see Part 3 and the *Parameters initialization* section for details):

```
1  params.timePerFrame = 15; params.maxTimeToProcess = 300;
2  params.maxDistToProcess = 150; params.stripSizeUm = 15;
3  params.spatialPartition = 3; params.timePartition = 4;
4  featuresNo = params.spatialPartition*params.timePartition;
```

To extract the high-dimensional feature representation of each experiment, we use:

```
1  featuresArray=zeros(featuresNo,length(experimentsLabels(:,1)));
2  for expIDX=1:length(experimentsLabels(:,1))
3      featuresArray(:, expIDX) =
   ↪    kymographToFeaturesVec(kymographsPaths(expIDX, :), measure,
   ↪    params);
4  end
```

To calculate the PCs, we use:

```
1  pcaResultsByMeasure = PCAOnAllExperimentsMeasurements(measure,
   ↪    params, {}, featuresArray);
```

To calculate the proportion of variance in the data attributed to each PC (here for PC1), we use:

```
1  pcIndex=1; pcaScores=pcaResultsByMeasure.(measure).score;
2  entire_data_variablity = var(pcaScores);
3  single_pc_explained_variance =
   ↪    var(pcaScores(:,pcIndex))/sum(entire_data_variablity);
```

Select k PCs, plot them with the corresponding labels:

```
1  pc1Scores=pcaScores(:, 1); pc2Scores=pcaScores(:, 2);
2  hold on; scatter(pc1Scores, pc2Scores);
3  title(sprintf('PC #1 about PC#2'));
4  xlabel('PC1'); ylabel('PC2');
5  c = cellstr(experimentsLabels); dx = 0.01; dy = 0.05;
6  text(pc1Scores+dx, pc2Scores+dy, c, 'Fontsize', 10,
   ↪    'Interpreter','none');
7  hold off;
```

8.4.6 Tips and Troubleshooting for Advanced Users

The pipeline was designed to be executed "as is", to analyze monolayer migration experiments. However, advanced users may want to customize components or tweak the pipeline. For these users we recommend to go through the documented code in *quantifyMonolayerMigrationMain.m*, and *quantifyMonolayerMigrationBulkMain.m*. We list common issues in ◘ Table 8.9, and errors in ◘ Table 8.10, that may arise while customizing the code and we give some suggestions how to handle them.

> **Take-Home Message**
>
> Our pipeline provides an analysis suite for monolayer migration experiments. It is designed for both users inexperienced in programming, as well as those more experienced users who wish to customize it further. The pipeline can be applied to extract traditional "wound healing" measures and/or more advanced spatiotemporal visualizations and qualifications. Its robustness was verified through experiments on data from multiple labs and cell systems (Gan et al., 2016; Zaritsky et al., 2015a, 2017b).

8

◘ **Table 8.9 Tips** for advanced users who wish to customize the code

Part	Topic	Tip
0	Parameters and directories initialization	When manually assigning parameters make sure to invoke the `initParamsDirs` function that will set the rest of the parameters to their default values, and create all necessary directories
1–2	Reusing intermediate results calculated in previous executions	The parameter `reuse` indicates whether to recalculate every output in the pipeline from scratch. When its value is set to 'true' (the default value) the pipeline reuses intermediate results from previous executions. All calculated intermediate outputs are saved for reuse in future executions

◻ **Table 8.10** **Common issues**, errors and warnings and how to solve them

Error	Probable cause	How to fix
File 'YourFilePath' nor images exist...	The file path to the timelapse image stack is wrong.	Carefully check your input, verify that you are using the folder separator character suitable for your operating system. You can also validate that the file exists by using MATLAB's built-in *exist* function.
Unrecognized function or variable 'params'...	Parameters structure not initiated.	Execute the 'Parameters Initialization' part-make sure that the function *initParamsDirs* is executed and assign its output to two variables named *params* and *dirs* (preserve order).
Unrecognized function or variable [FUNCTION NAME]...	Directories/workspacepath was not defined properly.	At the current folder pane, right-click on the MATLAB source code folder =>*Add to Path* =>*Selected Folders and Subfolders* or use MATLAB's *addpath* command.
mflimage file for frame No. ### was not found! Please run EstimateVeloctyFields	No velocity fields files found (###_mf.mat files) or no image files found.	Execute Part 1.

Solutions to the Exercises

✓ **Exercise 1**

The segmentation implemented for this pipeline is optimized for robustness with the goal to enable high-content automated analysis. This was achieved by relying on the assumption that the monolayer advances over time and never goes ''backwards'' (Part 1). This assumption significantly reduces the number of image patches that must be segmented as foreground or background at each time frame, since it enables to focus only on patches close to the previous segmentation. Importantly, our kymograph-based quantification is not sensitive to small deviations in the segmentation (Part 2), but is very sensitive to large segmentation errors, even if they occur only in a few frames. Thus, although other segmentation algorithms in the majority of the cases produce more accurate segmentation, we preferred robustness at the cost of reduced segmentation accuracy.

✅ Exercise 2

Run the following code snippet on each of the configurations, for both files *EXP_ 16HBE14o_1E_SAMPLE.tif* and *EXP_MDCK_HGFSF_1E_SAMPLE.tif*. The function `stepsScripts/KymograhpsByMeasure` computes and visualizes the kymographs. The visualization is rendered and saved using the custom function `utils/plotKymograph`, which is called from `KymographsByMeasure`.

```
1   params.patchSizeUm = patchSize; % i.e. 15 μm
2   params.pixelSize   =  pixelSize; % i.e. 0.879 μm
3   params.timePerFrame = timePerFrame; % i.e. 15 minutes
4   params.maxSpeed = maxSpeed; % i.e. 90
5   [params, dirs] = initParamsDirs(pathToFile, params);
6   calcSpatiotemporalRaw(params, dirs); % for velocity fields estimation
    ↪  + segmentation
7   allMeasuresToProcess = {'speed', 'directionality'};
8   KymographsByMeasure(params, dirs, allMeasuresToProcess);
```

✅ Exercise 3

When the `maxSpeed` parameter is set to a value below the true cells' maximal speed (e.g., $30\,\mu m*h^{-1}$), the estimated magnitude of the vector fields will be bounded by the underestimated search radius, leading to reduced velocities (as illustrated in ◘ Fig. 8.7, left). On the other hand, when the search radius is overestimated (by setting `maxSpeed` to a value higher than the true maximal speed) the potential cross-correlation matches grow quadratically leading to overestimation of the velocities magnitude (◘ Fig. 8.7, right). Increasing `patchSize` leads to a trade-off between having more information for the cross-correlation analysis at the cost of lower resolution in the segmentation and velocity granularity. See Part 1 for a thorough discussion.

✅ Exercise 4

For example, the wound healing rate may remain unchanged when a perturbation induces both increased directionality and impaired motility, canceling each other in the wound healing rate measurement. Another example is a perturbation that slightly reduces cells' speed, while enhancing front-to-back inter-cellular communication, together leading to unchanged wound healing rate. The latter phenotypes were reported for RHOA-inhibited cells in Zaritsky et al. (2017b) and were first identified by visualizing the spatiotemporal dynamics using kymographs (◘ Fig. 8.10).

✅ Exercise 5

Lines #4-#10 in the code below enable a simple visualization of the kymographs. An alternative is to use the function `utils/plotKymograph` to visualize the kymograph.

```
1   speedKymo1 = load(speedKymo1Path).speedKymograph; speedKymo2 =
    ↪   load(speedKymo2Path).speedKymograph;
2   maxTimeToCompare = 50; % to enforce identical dimensions
3   subtractionResult = speedKymo1(:, 1:maxTimeToCompare) - speedKymo2(:,
    ↪   1:maxTimeToCompare);
4   h = figure;
5   hold on;
6   colormap('jet');
7   imagescnan(subtractionResult);
8   haxes = get(h,'CurrentAxes');
9   set(h,'Color','w');
10  hold off;
```

An example of results of this procedure can be seen in ◘ Fig. 8.8; the *Difference* kymograph is the result of subtracting two speed kymographs.

Acknowledgements   We thank Andres Nevarez for providing feedback and for testing the pipeline. We would like to thank Simon F. Nørrelykke and Kota Miura for providing insights that improved this chapter considerably.

References

Caldas VE, Punter CM, Ghodke H, Robinson A, van Oijen AM (2015) iSBatch: a batch-processing platform for data analysis and exploration of live-cell single-molecule microscopy images and other hierarchical datasets. Mol Biosyst 11(10):2699–2708

Candès E, Demanet L, Donoho D, Ying L (2006) Fast discrete curvelet transforms. Multiscale Model Simul 5(3):861–899. https://doi.org/10.1137/05064182X

Carpenter AE, Jones TR, Lamprecht MR, Clarke C, Kang IH, Friman O, Guertin DA, Chang JH, Lindquist RA, Moffat J, Golland P, Sabatini DM (2006) Cell Profiler: image analysis software for identifying and quantifying cell phenotypes. Genome Biol 7(10):R100

Deforet M, Parrini MC, Petitjean L, Biondini M, Buguin A, Camonis J, Silberean P (2012) Automated velocity mapping of migrating cell populations (AVeMap). Nat Methods 9(11):1081–1083

Gan Z, Ding L, Burckhardt CJ, Lowery J, Zaritsky A, Sitterley K, Mota A, Costigliola N, Starker CG, Voytas DF, Tytell J, Goldman RD, Danuser G (2016) Vimentin intermediate filaments template microtubule networks to enhance persistence in cell polarity and directed migration. Cell Syst 3(3):252–263

Geback T, Schulz MM, Koumoutsakos P, Detmar M (2009) TScratch: a novel and simple software tool for automated analysis of monolayer wound healing assays. BioTechniques 46(4):265–274

Jonkman JE, Cathcart JA, Xu F, Bartolini ME, Amon JE, Stevens KM, Colarusso P (2014) An introduction to the wound healing assay using live-cell microscopy. Cell Adh Migr 8(5):440–451

Lee RM, Kelley DH, Nordstrom KN, Ouellette NT, Losert W (2013) Quantifying stretching and rearrangement in epithelial sheet migration. New J Phys 15(2):025036

Liang CC, Park AY, Guan JL (2007) In vitro scratch assay: a convenient and inexpensive method for analysis of cell migration in vitro. Nat Protoc 2(2):329–333

Masuzzo P, Van Troys M, Ampe C, Martens L (2016) Taking aim at moving targets in computational cell migration. Trends Cell Biol 26(2):88–110

Milde F, Franco D, Ferrari A, Kurtcuoglu V, Poulikakos D, Koumoutsakos P (2012) Cell Image Velocimetry (CIV): boosting the automated quantification of cell migration in wound healing assays. Integr Biol (Camb) 4(11):1437–1447

Ng MR, Besser A, Danuser G, Brugge JS (2012) Substrate stiffness regulates cadherin-dependent collective migration through myosin-II contractility. J Cell Biol 199(3):545–563

pca (2020) Xpca. https://www.mathworks.com/help/stats/pca.html

Pearson K (1901) LIII. On lines and planes of closest fit to systems of points in space. Lond Edinb Dublin Philos Mag J Sci 2(11):559–572. https://doi.org/10.1080/14786440109462720

Santiago JG, Wereley ST, Meinhart CD, Beebe D, Adrian RJ (1998) A particle image velocimetry system for microfluidics. Exp Fluids 25(4):316–319

Simpson KJ, Selfors LM, Bui J, Reynolds A, Leake D, Khvorova A, Brugge JS (2008) Identification of genes that regulate epithelial cell migration using an siRNA screening approach. Nat Cell Biol 10(9):1027–1038

Slater B, Londono C, McGuigan AP (2013) An algorithm to quantify correlated collective cell migration behavior. BioTechniques 54(2):87–92

Suarez-Arnedo A, Figueroa FT, Clavijo C, Arbeláez P, Cruz JC, Muñoz-Camargo C (2020) An image j plugin for the high throughput image analysis of in vitro scratch wound healing assays. PLoS ONE 15(7): e0232565. https://doi.org/10.1101/2020.04.20.050831

Vitorino P, Meyer T (2008) Modular control of endothelial sheet migration. Genes Dev 22(23):3268–3281

Zabary Y, Zaritsky A (2020) Quantifying monolayer cell migration sample dataset. https://doi.org/10.5281/zenodo.4308385

Zaritsky A, Natan S, Horev J, Hecht I, Wolf L, Ben-Jacob E, Tsarfaty I (2011) Cell motility dynamics: a novel segmentation algorithm to quantify multi-cellular bright field microscopy images. PLoS ONE 6(11):e27593

Zaritsky A, Natan S, Ben-Jacob E, Tsarfaty I (2012) Emergence of HGF/SF-induced coordinated cellular motility. PLoS ONE 7(9):e44671

Zaritsky A, Manor N, Wolf L, Ben-Jacob E, Tsarfaty I (2013) Benchmark for multi-cellular segmentation of bright field microscopy images. BMC Bioinf 14:319. https://doi.org/10.1186/1471-2105-14-319

Zaritsky A, Kaplan D, Hecht I, Natan S, Wolf L, Gov NS, Ben-Jacob E, Tsarfaty I (2014) Propagating waves of directionality and coordination orchestrate collective cell migration. PLoS Comput Biol 10(7):e1003747

Zaritsky A, Natan S, Kaplan D, Ben-Jacob E, Tsarfaty I (2015a) Live time-lapse dataset of in vitro wound healing experiments. Gigascience 4:8

Zaritsky A, Welf ES, Tseng YY, Angeles Rabadán M, Serra-Picamal X, Trepat X, Danuser G (2015b) Seeds of locally aligned motion and stress coordinate a collective cell migration. Biophys J 109(12):2492–2500

Zaritsky A, Obolski U, Gan Z, Reis CR, Kadlecova Z, Du Y, Schmid SL, Danuser G (2017a) Decoupling global biases and local interactions between cell biological variables. Elife 6:e22323

Zaritsky A, Tseng YY, Rabadán MA, Krishna S, Overholtzer M, Danuser G, Hall A (2017b) Diverse roles of guanine nucleotide exchange factors in regulating collective cell migration. J Cell Biol 216(6):1543–1556

Zhou FY, Ruiz-Puig C, Owen RP, White MJ, Rittscher J, Lu X (2019) Motion sensing superpixels (MOSES) is a systematic computational framework to quantify and discover cellular motion phenotypes. Elife 8:e40162

Supplementary Information

K. Miura, N. Sladoje (eds.), *Bioimage Data Analysis Workflows–Advanced Components and Methods,*
Learning Materials in Biosciences, https://doi.org/10.1007/978-3-030-76394-7

Index

A

Adaptive moment estimation (Adam)   72, 81
Anaconda   31–35
Analyze Particles command   101
Application programming interface (API)   96
Artificial intelligence (AI)   61
Artificial neural network (ANN)   61
Auto-threshold method   25

B

Batch Processor   12, 13
Benchmarking   108–111
Binary cross-entropy (BCE)   72
Binary threshold   25
Bioimage analysis deconstruction
– components   119
– dataset   123
– description   119–120
– LSM-W2   122
– Merryproj   122
– MorphoGraphX (MGX)   122
– Smooth 2D manifold   122
– SurfaceProject   122
– SurfCut (*see* SurfCut)
– tools   123
– workflow   119
  – benchmarking   140–143
  – code refactoring   133–137
  – FibrilTool   144
  – graphical scheme   125–126
  – identification of components in code 130–133
  – identification of components in textual description   124–125
  – prerequisites and limitations   126–130
  – shiftmask in Z-Axis direction   137–140
Bland-altman analysis   107–108
Bokeh   51–54

C

Categorical cross-entropy (CCE)   72
Cell 122 Tracking Challenge (CTC)   63
CLIJ   90
– application programming interface   96
– *clear( )* command   95
– functionality   94
– hardware   96–97
– ImageJ macro   94
– ImageJ's built-in *getTitle( )* command   94
– installation   93
– macro processes   93
– mean filter   93
– *nVendor Awesome Intelligent,*   94
– processing time   93
– *pull( )* command   95
– result images   95
– toolbox   95
– website and API reference   99
CLIJ2   93
CLIJx   95
Collection, definition   2
Command-line headless methods   19–21
Command Recorder   23, 24
Comma-separated values (CSV)   79
Component, definition   2
Computer Vision   62
Crowdedness
– calculation   163
– description   160–161
– MTPV   161–163
Cutting-edge algorithms   3

D

Data augmentation (DA)   82–84
Data collecting macro   31
Data storage array   21, 22
DeepImageJ bundled model   76–79,79
Deep learning (DL)
– annotation process   62
– Computer Vision   62
– CTC   63
– data augmentation   82–84
– data, download and split   65–66
– DeepImageJ bundled model   76–79
– Google Colab notebook set up   64–65
– graphics processing unit   62
– Halo and receptive field of network   81–82
– hyper-parameters   80–81
– images processing using DeepImageJ and MorpholibJ   79–80
– International Symposium on Biomedical Imaging (ISBI)   62
– optimizers   91
– tools and software packages   64
– training data   63
– U-Net   61, 62
  – background mask   66
  – cell mask   66
  – convolutional neural network   69–72
  – loss and accuracy measures   72–73
  – preparing the data for training   67–68
  – semantic segmentation   67
  – training schedule   73–75
Dice coefficient   72

Printed in the United States
by Baker & Taylor Publisher Services